EASY-PC Handbook

EASY-PC Handbook

IAN SINCLAIR

NEWNES

Newnes
An imprint of Butterworth-Heinemann Ltd
Linacre House, Jordan Hill, Oxford OX2 8DP

A member of the Reed Elsevier plc group

OXFORD LONDON BOSTON
MUNICH NEW DELHI SINGAPORE SYDNEY
TOKYO TORONTO WELLINGTON

First published 1995

British Library Cataloguing in Publication Data
A catalogue record for this book is
available from the British Library

ISBN 0 7506 2281 4

Library of Congress Cataloguing in Publication Data
A catalogue record for this book is
available from the Library of Congress

Typeset by Butford Technical Publishing, Bodenham, Hereford
Printed and bound in Great Britain by Clays Ltd, St Ives plc

Contents

Preface

Drawing a circuit diagram and subsequently planning and drawing the silk-screen outlines and the copper track plan for a printed circuit board are activities that are rightly regarded as monotonous and liable to error when they are done by the traditional methods. For a considerable time it has been possible for the larger manufacturer to use computers, mainframe or mini, to aid these actions, but at a cost that would be unacceptable for the small-scale designer or the educational user. The standardization of the PC type of machine and the remarkable amount of computing power that modern examples (using the 80386 or 80486 chips) can achieve now bring computer-aided drawing and PCB planning methods within the reach of virtually any user.

The program that makes such an advance possible is called EASY-PC, from Number One Systems, and it exists in two forms. The basic form allows you to draw circuit diagrams, linear or digital, using a set of ready-made symbol shapes that you can manipulate on screen and align with lines that are maintained straight and drawn horizontally, vertically or at 45°. You can, of course, add to the symbol library for yourself. The PCB layout can be drawn in the same way, using shapes that are the exact size of electronic components, along with pads and track lines. Though the PCB layout has to be drawn manually, you can manipulate it as much as you need before printing so as to ensure that the printout of the silk-screen view and of the copper track pattern will be perfect. Up to two silk-screens and eight copper layers can be catered for.

The Professional version of EASY-PC allows the PCB connections to be drawn *automatically* after the circuit diagram has been drawn, and you can then arrange the components on the circuit board and re-route crossing tracks until the layout is satisfactory, with no need to specify each connection. The program can detect unwanted track crossings and points where there might be a danger of tracks shorting. In addition, when the schematic (circuit diagram) has been drawn using EASY-PC Professional, you can analyse the circuit action. Linear circuit analysis is carried out by the Analyser software, providing graphs of amplitude and phase plotted against frequency, and digital circuit analysis is carried out by the Pulsar software, providing simulated pulse inputs and showing, as if on a multi-track oscilloscope, output pulses. EASY-PC Professional contains both the Analyser and Pulsar software, limited only in the number of components they can cope with, and you can upgrade if you need to analyse more elaborate circuits. Keep a close watch on the electronics press (or have your name added to the Number One Systems mailing list) to see upgrade offers.

This, along with the provision (also in EASY-PC) for multi-layer boards and file output to Gerber Photo-plotting machines and to numerically-controlled drilling machines, makes the professional version more suited for the designer of elaborate circuits who needs to design a large number of boards. Nevertheless, substantial numbers of hobby users of EASY-PC upgrade to the Professional version, and a brief description of this version is included in this book.

The purpose of this book is to introduce the use of EASY-PC to the small-scale user of PC boards who has not realized the amount of assistance that a computer of the PC type can provide in this respect. A reasonable knowledge of electronics will be assumed, but little of computing, so that the use of the software will be described in much more detail than it would be in a book aimed at the more seasoned computer user. Since EASY-PC is also likely to be used by many who have never laid out a circuit in a PCB, the descriptions and the examples are deliberately made simple. There are more elaborate examples included with the EASY-PC package.

Chapter 6 consists of an alphabetical listing of EASY-PC actions with their active keys and description. This has been done because it would take too long to provide an illustrated example of each. The illustrations cover the main actions, and the range of these is sufficient to ensure that other actions never look entirely unfamiliar because they will follow the pattern of the established actions. Where an action has not already been illustrated (and needs to be) there is an illustration in this section, and there are several illustrations that duplicate information simply so that you do not need to thumb through the book looking for information.

The order of description of actions is not that followed in the manual. The emphasis in this book is on the creation of schematics first, followed by the translation of schematic to PCB layout. This is not the order followed in the manual, and the reason for this is that most readers will be more familiar with schematics than with PCB layout. By dealing with the more familiar action first, I feel that you will be better able to concentrate on the features of EASY-PC. In addition, when you upgrade to EASY-PC Professional, the creation of the schematic is all-important, enabling you automatically to analyse the circuit action and to create a PCB layout (which can then be edited) without the need for further drawing.

NOTE: Throughout this book, the term *schematic* will be used to refer to a circuit diagram, in line with the EASY-PC manual.

Ian Sinclair
Spring 1995

1 Machine and program

Your computer

The type of machine that we now describe as a PC means one that is modelled on the IBM PC type of machine that first appeared in 1980. The reason that this type of machine has become dominant is the simple one of continuity – programs that will work on the original IBM PC machine will also work on later versions and will still work on today's much more advanced machines. By maintaining compatibility, the designers have ensured that when you change computer, keeping to a PC type of machine, you do not need to change software (programs). Since the value of your software is much greater than the value of the hardware (the computer itself), this has ensured that the PC type of machine has become dominant in business and other serious applications.

Other machines, such as Apple, Atari, Commodore Amiga, Acorn Archimedes and the Amstrad PCW and Notepad machines, are not compatible with the PC nor with each other. They also have less choice of software and more expensive components, though recent price changes have made the Apple Mac machines much more serious competitors. Compatibility of software on PC machines is not perfect, however, and some software that is being written now (identified as *Windows* programs) will not run on the older machines, though a lot of programs, including EASY-PC, still do. EASY-PC will run on any variety of PC machine, including the types known by the processor types as 8088, 8086, 80286, 80386 and 80486, or by the letters XT or AT.

In short, a real PC machine is currently identified by the following points:

1 It uses a microprocessor which is of the Intel type, though not necessarily manufactured by Intel. This microprocessor will be of a type number that starts with 80 and ends in 86 – the 8086 and 80286 are very early types, and you should preferably use a machine containing the 80386 or 80486. The new Pentium chip is also of this family.

2 It uses a program called MS-DOS or PC-DOS as a master controlling system (an operating system), to enable it to load and run all other programs.

A machine satisfying these requirements can carry a well-known name or, by contrast, it might be produced by a back-street small-scale assembler, or it can be home-made. The standardization of the machine is such that it is outstandingly easy to assemble and the quality of the machine is not necessarily related to the price that you are asked to pay for it.

Because the PC machine uses open architecture, meaning that anyone can produce circuit boards for it, there have been many variations in the past that have resulted in differing standards, particularly for the video output. Modern machines have standardized on a video system called *VGA*, which permits the use of 640 x 480 graphics displays, meaning that displays can consist of 640 dots horizontally and 480 vertically on the monitor screen. If you are buying or assembling a machine you should insist on this standard. Many VGA circuit boards also allow higher resolutions such as 800 x 600 to be used, but EASY-PC does not require such high resolutions and it can even be used on the older CGA or Hercules display systems if you are using an old machine. You should not buy a system that uses these older systems, however (even if you can find one by now), but if you are using an old machine you do not need to replace it in order to run EASY-PC.

- The simple version of EASY-PC, used along with a VGA screen, produces screen views in which circles are vertically elongated into ellipses, as several of the screenshots in this book illustrate. The shape is perfect on a printed copy, and this distortion of view does not affect your use of the program – in fact it allows for faster operation if your computer is an old and slow one. EASY-PC Professional displays perfect round circles on a VGA screen.

A printer is essential to print out the schematics and PCB diagrams. The three main printer types for such purposes are dot-matrix impact printers, laser printers and ink-jet printers, and EASY-PC is designed to make effective use of any good printer in any of these classes. Very old 9-pin dot-matrix printers are liable to be less satisfactory, but the only printers that are definitely impossible to use are the daisywheel types, once used for office purposes. EASY-PC can also provide output to pen-plotters.

The other crucial requirements are memory and disk space. No PC machine produced in the last eight years has contained less than 640 Kbyte of available read-write memory (RAM) and modern software demands this as a minimum requirement. Most modern machines are fitted with a minimum of 2 Mbyte (2048 Kbyte) of RAM, and to run the more elaborate modern programs that use the Windows system machines need 4 Mbyte or more of such memory. EASY-PC in its basic version needs at least 512 Kbyte of RAM (Random Access Memory, meaning memory that can be read or written), an amount that any modern machine can supply. The professional version requires at least 640 Kbyte of RAM, preferably more. Note in this respect that one byte of memory corresponds to one unit of data (one printed character, for example) and that 1 Kbyte = 1024 bytes, 1 Mbyte = 1024 Kbyte.

Older machines often relied on using programs stored and distributed on floppy disks. As programs have become more elaborate, the use of a permanent (hard) drive has become essential, and modern machines use a hard drive of 80 Mbyte as an absolute minimum requirement. The simpler version of EASY-PC can be run from a floppy disk, but this is not ideal, and a hard drive is strongly recommended. If you need to use the professional version of EASY-PC your machine should use 1 Mbyte or more of RAM and a hard drive of adequate capacity for this and all the other programs you intend to use. The current advice concerning RAM and disk capacity is to have as much as you can afford – and preferably more.

An option that was once rare but which is now almost universal is the mouse. This is a small trolley which, when controlled by a program (called MOUSE.COM) that stays resident in the memory, will control the position of a marker (a cursor) on the screen. The mouse can be used in drawing and for selecting and moving objects that appear on the screen. The mouse is essential to many modern programs, and though its use for EASY-PC is optional, it is very strongly recommended because methods that depend on using keys are clumsy by comparison. Note that the mouse cannot be used unless its controlling program is installed each time the computer is switched on. See later in this chapter for details of how to install and use a mouse.

Introducing EASY-PC principles

EASY-PC is a program of the CAD (computer-aided design) type, meaning that it is intended to make it easy for you to draw schematics and PCB layouts on screen, avoiding the tasks of rubbing out and re-drawing that plague you if you work directly on paper. Chapter 7 deals separately with the professional version, which will analyse the action of a circuit, working from the schematic, and also develop the PCB layout from the schematic in a semi-automatic way; but since the two programs are almost identical in other respects, this book is concerned mainly with the simpler version, which is intended for the small-scale user. You can upgrade from EASY-PC to the professional version at any time by paying the current price difference between the programs – watch out for special offers from time to time.

In addition to producing PCB layouts, EASY-PC can also be used for producing schematics (circuit diagrams) of professional quality, eliminating the need for the tedious work of using dry transfers or stencils for this type of work. As always, the advantage of using the computer for this type of work is that you need never repeat any action, because you can keep a file of each drawing and use an old drawing as the basis of a new one, changing only the portions that need to be changed. You can also keep units, such as power supply units, as a separate file that can be used over and over again in drawings.

EASY-PC in its PCB drawing mode allows you to work with multi-layer boards, of up to eight layers, with a silk-screen for front and back. The board size can be up to 17" x 17" (430 mm x 430 mm), which covers virtually all the boards that the small-scale user is likely to require.

On the board, EASY-PC will cater for up to 1,500 ICs, up to 5,000 tracks (with up to 12,000 segments), and up to 4,000 pads (other than the pads used in the ICs). On a schematic, a maximum of 100 different symbols and 6,000 characters of text can be placed in one diagram.

Getting back to the PCB layout, you can work with a maximum of eight different widths of track on a single board, selecting from a possible 128 widths ranging from 0.002" (0.05 mm) to just over 0.5" (13 mm). You can use as many as sixteen different sizes of pads, again selected from 128 possible sizes that range from the same choice as for track widths. The pads can be circular, oval, square, rectangular or they can be finger-shaped for edge connectors.

Given this array of items, what can you do with them? You can place a track on the screen view, and change the width and the route as much as you like. You can create arcs and other shapes of curved track. You can place a grid display on the screen to make it easier to guide your positioning of tracks and components, and you can also use a snap action which will ensure that a track or component that you place near a grid line will be positioned exactly on that grid line. You can also place pads and tracks off the grid if you want to.

The usual advantages of using the computer apply. You can copy a piece of layout, so making it much easier to construct layouts in which a shape is repeated (such as bus connections). You can also rotate shapes and mirror them, another useful pair of ways of reducing the effort of layout drawing. You can work with lines that join at 45° or 90° only, or you can opt to have lines joining at any angle you want to use.

Preliminaries

NOTE: There is no space in this book to discuss in detail the MS-DOS operating system of the computer, and I assume that you know that each command to the operating system is typed as a word or phrase and followed by pressing the ENTER or RETURN key. This key is usually marked with a hooked arrow, Figure 1.1, and since it is referred to as the

Figure 1.1 The RETURN or ENTER key marking: this arrow shape is used in place of lettering and the key that is marked ENTER, on the right-hand corner of the keyboard, carries out the same action.

RETURN key in the EASY-PC manual the same name will be used throughout this book. Other software may refer to this key as the ENTER key. If you want an easy introduction to MS-DOS use, take a look at *MS-DOS Made Simple* from Butterworth-Heinemann.

The program as it is supplied will be on a single 3½" disk or, to special order only, on two 5¼" disks. All modern machines use the 3½" disk type, because these hold more data than the older 5¼" variety and are better protected against accidental damage. The files on the supplied disk(s) are compressed so as to take up less disk space, so that you cannot simply copy them to the hard drive or to another floppy disk. They cannot be used until they have been expanded again, an action that is performed by the installation routines of EASY-PC.

- The 5¼" disk option is still available, but it cannot be guaranteed indefinitely now that all PC machines use the 3½" format.

Remember that you should not under any circumstances use the disk that you received with EASY-PC for working with the program. The correct procedure is to install the program files on the hard drive or on another floppy disk and then use this working copy – for safety it is better to hold at least one floppy disk of working files as a backup. Hard drives are reliable but their life, like that of any mechanical component, is limited, and at some time a hard drive will fail completely (though most computers will have been replaced by a more modern design before the hard drive fails). This is when the value of backup disks becomes obvious, because it allows all of your programs to be copied to a new hard drive.

The use of backups also becomes important if the working copy of a program becomes corrupted for any reason. This is not common, but odd things can happen, and users have been known to delete files by mistake after several hours at the keyboard. If you have a backup you can restore your working copy on the hard drive or replace a working copy on a floppy. Without a backup you need to re-install from scratch, and if anything has happened to corrupt the files on the only master copy that you bought you have no option but to buy another copy.

Remember that backup copies are for your use only as a safeguard against the loss of an important program. You must not give away or lend copies to anyone else, as this would be a breach of copyright and of the agreement that you have implicitly given when you opened the envelope containing the distribution disk(s). You should be equally reluctant to receive dubious copies of software; apart from moral considerations (are you receiving stolen goods?) such disks are often a source of a computer virus infection, which could plague you for a long time and possibly make it difficult to use the computer.

Installing the program – backup

The machine should have been switched on, with the usual screen prompt (either C:\ ,if you have a hard drive, or A:\ for a machine that uses only floppy disks) showing. Before you start installation, you should make a backup copy of the distribution disk(s). If you are using a 3½" drive, prepare a formatted blank 3½" disk, and if your machine uses a 5¼" drive prepare two blank formatted disks of this size. Label the disk(s) as *Backup of EASY-PC* (number them 1 and 2 if you are using 5¼" disks). You need to know where the essential MS-DOS files are located on your computer. For a floppy-only machine they will be on a separate system floppy. On a hard drive machine, they will be either on the root directory (C:\) or in a special directory called C:\MSDOS or C:\DOS.

For a machine using a hard drive C:

1 Ensure that you are using the directory that contains the MS-DOS files. For modern machines this usually means the C:\MSDOS or C:\DOS directory. The MS-DOS command to log on to this directory is `CD \MSDOS` or `CD \DOS` – you will have to determine for yourself which name is used.

2 Type `DISKCOPY A: A:` and press the RETURN key. You will see the message:

```
Insert the source disk in the A: drive
```

meaning that you should place the distribution disk (the first one if you use two), in the A drive. Do this and press the RETURN key.

3 After some time you will see the message:

```
Insert the destination disk in the A: drive
```

which is a signal to remove the distribution disk and to insert the blank 3½" disk or the first of the 5¼" disks.

4 You may need to swap disks several times, depending on the memory size. If you are using MS-DOS 6.2 (or later) this will not be needed, but earlier versions store the data temporarily in memory, limiting the amount that can be transferred at one time. You will be prompted by the messages on the screen as to which disk to insert.

5 Write-protect the original disk(s) and store in a cool dry place well away from sunlight or magnets.

6 You can make another backup set, just to be certain. You should use only the copies for running EASY-PC if you work from floppy disks. If you use a hard drive, one backup set (which will not normally be used) is enough.

For a machine using two floppy drives:

If you are using a machine with no hard drive but two floppy drives, and you want to copy from the distribution disk in one drive to a blank formatted disk in the other, the two floppy drives must be identical – either both 5¼" (very unusual nowadays) or both 3½". You cannot use the DISKCOPY program with drives that are not identical. The procedure for such a machine is:

1 Place an MS-DOS system disk in drive A and type `DISKCOPY A: B:` (press RETURN).

2 When you are prompted, place the distribution disk (or the first of two) in the A drive and press the RETURN key again.

3 Do not remove any disks until the action is complete – you will see the message appear.

Remember that you can make more than one backup of the distribution disk(s), but at this stage it is more important to get on with creating a working set of files.

Installation – main files

Like most modern programs, EASY-PC can be installed by using a program that carries out the hard work for you. All you need to do is to answer questions or make choices as they appear on the screen. The same procedure is used whether you install on a hard drive (C) or on a second floppy disk (using a machine which has two drives labelled as A and B). Installation is done using the original 3½" (distribution) disk as follows – if you use 5¼" drives there will be two distribution disks and you will be asked to change disks midway through installation.

1 Insert the distribution disk (the first if you have two) in drive A. If you use a machine with no hard drive but two floppy drives, insert another blank formatted disk in drive B.

2 Make sure that the A drive is being used (if in doubt, type `A:\` and press the RETURN key).

3 Type `INSTALL` and press the RETURN key. This will start the installation process, unless you have inserted the wrong disk of a two-disk set.

4 You will be asked to confirm that you have a backup copy of the distribution disk. If you did not make one earlier, you can escape from installation at this point to make a backup.

5 You will be asked to select by number whether you are installing to the hard drive, to a 3½" floppy or to a pair of 5¼" low-density disks. You can also opt to escape if you do not have disks ready.

6 You are asked to select the drive letter for installation, with the default of C. To use the hard drive option, simply press RETURN, otherwise type a drive letter such as A or B.

7 For hard drive installation, you are asked to select the directory to use. The default is to create a directory called EASYPC, and you can make this selection by pressing the RETURN key. The other options are to use the current hard drive directory, whatever that happens to be, or create a directory called PCB, or escape from installation. Use the default of EASYPC unless you already have a directory of that name.

8 You are asked to select the type of video system that your computer uses. The normal default is VGA, and the program will guide you on what it can detect automatically.

9 You are asked what type of dot-matrix printer you will use, selecting from None, 9-pin 80-column, 9-pin 132-column, 24-pin 80-column and 24-pin 132-column. You will also be asked if you are using a laser printer or a pen-plotter and asked for further information on the type.

10 You are asked about the mouse that your computer uses. The options are None, a 2-button type or a 3-button type. The 2-button type (Microsoft mouse) is much more common, and it is likely that this is the type that your computer is using even if the mouse has three buttons. If your mouse is a three-button type and you have a suitable software driver (such as the Mouse Systems driver) the middle button carries out the Pan action (see later). If your mouse has three buttons but you are using the normal (Microsoft) driver for a two-button

mouse then the central button has no effect; the left and right buttons work normally.

11 You will be asked if you want in install the Gerber output program – use this only if you intend to make use of a PCB preparation bureau which has the associated equipment. You will also be asked about the Drill output program and again you answer YES only if you intend to send disks to a bureau which will have the corresponding numerically-controlled drill hardware.

12 You are asked about which standard IC libraries you want to install, such as the 7400 TTL set, and also about the example files. You should opt to install all libraries and examples at this stage.

13 When you have made all the selections that are required, you will be told that installation is complete. How you run EASY-PC from this point onwards now depends on whether the installation was made to a hard drive directory or to another floppy disk.

To use EASY-PC from a floppy, place the floppy in drive A. Make sure that the prompt message on the screen is A: and type `EASYPC` (press RETURN). The program will then run. You should use the other floppy drive to contain data files generated from EASY-PC.

To use EASY-PC from the hard drive directory, first ensure that you are using the correct hard drive directory. If the program was installed in the C:\EASYPC directory, type `CD EASYPC` (press RETURN). If the program was installed into a directory called, for example, C:\PCB then type `CD PCB` (press RETURN). Check after the CD (change directory) command that your directory contains the EASY-PC files by typing `DIR` (press RETURN) and reading the display of files.

When you see that you have the correct directory, type `EASYPC` and press the RETURN key. This will start the program.

- If you know how to create and use a batch file you can use this much more convenient way of locating the directory and starting EASY-PC. See any good book on MS-DOS for details of how to create a batch file.

Keyboards

The original form of IBM PC keyboard used 83 keys laid out with the function keys (marked as F1 to F9) grouped at the left-hand side and the number keypad at the right. The number keys on the right also doubled as cursor keys with a number lock key used to avoid the need to hold down the SHIFT key when numbers were being entered from this keypad. This layout was widely criticized, and some of the points that were made were

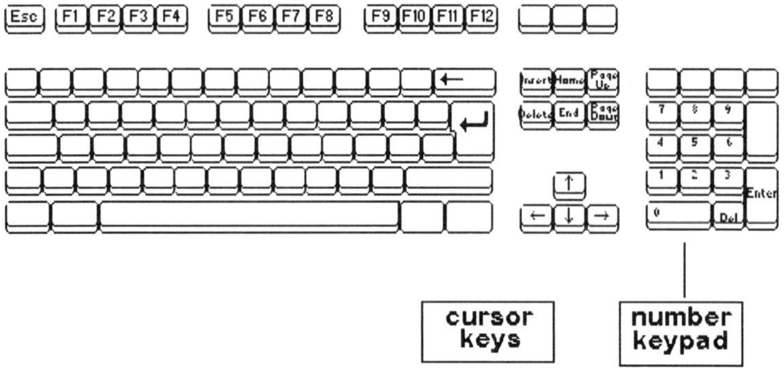

Figure 1.2 The modern 102-key keyboard layout, showing some of the important keys.

that the ESC key was too close to others, there was only a single CTRL key at the left-hand side, and the combining of the cursor keys with the number keypad was inconvenient.

The later type of keyboard, almost universal nowadays, has 102 keys, which are laid out in more logical groupings (Figure 1.2). The function keys are set across the top of the keyboard, with the ESC key isolated to the left. The cursor movement keys are now separated from the number keypad, as also are the INSERT, HOME, PageUp, PageDn, DELETE and END keys. The CTRL keys are duplicated so that they can be used with either hand, making it easier to press CTRL with any other key. This is the type of keyboard that any modern machine will use.

Note that the EASY-PC manual refers to using the number keys at the top of the keyboard. This is not necessary if you are using a modern keyboard and the computer is set up (using the CMOS-RAM setup) with the NUM LOCK on by default – machines are nearly always set up in this way now. The keypad keys will then provide the numbers and the four cursor keys, Figure 1.2, will provide for moving the cursor around the screen if you have no mouse installed (but you cannot move diagonally as you can using a mouse).

Speed of use

The speed at which EASY-PC operates depends very much on your computer. The slowest speed is enforced if you have to work from a floppy disk and if you are using an old design of machine (though they were still being manufactured in 1991, some ten years after the original appeared) with the 8088 or 8086 processor. The fastest possible speed will be obtained if you are using a modern 80386 or 80486 machine with a hard drive and a fast clock speed of 33 MHz upwards. Speed is not essential when you are typing information into the machine, because your typing speed is then the determining factor.

However, when the machine has to work out positions of objects and calculate the lengths of lines, it needs as much processing speed as it can obtain, so that it will benefit from a fast processor and a high clock rate. The use of a co-processor does not affect EASY-PC.

- The part of EASY-PC which is most critical for speed is printing, because so much data has to be assembled and so many calculations made. If you work from Windows you may be able to carry out printing as a background action, allowing you to carry on with other Windows programs while you are printing.

The disk drive speed becomes important when a lot of symbols and shapes need to be read from a library held on the disk. Reading and writing a hard drive is much faster than the corresponding actions on a floppy disk, because the hard drive is spinning all the time that the computer is working, and spins at ten times the speed of a floppy drive. On modern machines, the rate of reading from the hard drive can be greatly increased by using a memory cache, such as SmartDrive (part of the MS-DOS system) and by using a local bus design of computer.

Tailpiece – using a mouse

The mouse is another way in which information is fed into the computer, and for graphics and drawing (CAD) programs, along with many DTP programs, it is the only really serious way of working with such programs. It is also essential for using Windows. The mouse consists of a casing containing a heavy ball which can be rotated when the mouse, Figure 1.3, is moved. The movement of the ball is transmitted to small rollers and then sensed in various ways, depending on the make of the mouse, such as magnetic changes, or by light reflection, and the signals that indicate the movement are returned to the computer to be used in moving the cursor.

The mouse action depends on software being present, and this can cause a considerable amount of confusion. The program MOUSE.COM has to be run in order to make a standard Microsoft mouse usable with programs that have been designed for mouse action. This does not, however, guarantee that the mouse will work as a cursor mover when you are using your favourite word processor, spreadsheet or database program. If you have bought and fitted a mouse independently of the manufacturer of the computer, follow closely the instructions that come with the mouse, and check

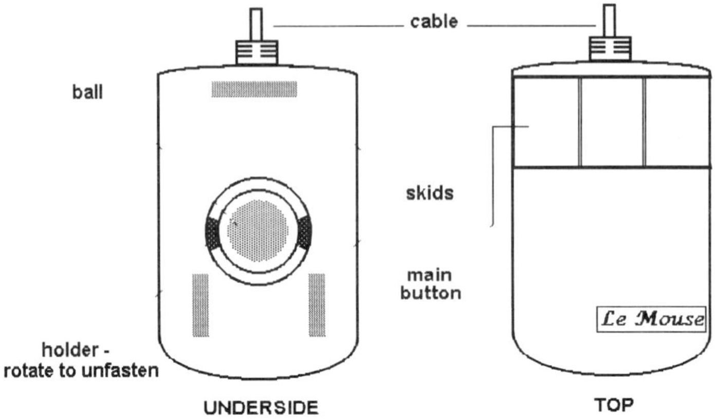

Figure 1.3 The mouse and its detachable parts. A software driver must be loaded in order to operate the mouse.

that the mouse is the correct type for your computer. EASY-PC will work with either the Microsoft or the Mouse Systems type of mouse. If you use the Windows system, the mouse will work along with all programs that you run using Windows.

Few computers allow for a mouse to be connected (to a mouse port) and the usual option is to use a serial mouse which plugs into a serial port. The advantage of using a serial mouse is that it is more likely to be possible to connect a serial mouse to any variety of PC machine than the Bus Mouse, because virtually any PC will have a serial port. The snag is that you may very well need the serial port for other actions, such as the use of a modem or for connections between computers. Many machines now provide more than one serial port, allowing both mouse and modem to be used.

Whatever mouse is used, it will have no effect unless the MOUSE.COM program is run at the time the computer is started up (booted). This is ensured by editing the file called AUTOEXEC.BAT and placing in the file a line such as:

```
C:\MSDOS\MOUSE
```

to run the mouse driver software. See any good text book on MS-DOS, such as the *MS-DOS 6.0 Pocket Book* for details of the AUTOEXEC.BAT file.

Mouse methods

The mouse cursor which appears on the screen when Windows is in use is shaped like an arrow, but the shape will change, depending on how the mouse is being used, when Windows programs are running. The cursor shape does not change while you are running EASY-PC. This cursor provides the main control actions. Three actions in particular are important: clicking, double-clicking and dragging.

Clicking is used to select an item that you want by placing the cursor over the item name or icon (the image next to a name) and then depressing the left-hand mouse button and releasing it, fairly quickly. If the mouse cursor is over a filename or its icon at the time, the name will change colour (or shading) to show that it has been selected for one of the actions of Windows or whatever program is in use. If the cursor is over the name of a menu (top line of screen) that menu will be displayed.

Double-clicking means pressing and releasing the left-hand mouse button twice in quick succession. If the mouse arrow is over a program name at this time, the program will run. If the cursor is over an item on a menu, that item will be chosen and used.

- If you double-click too slowly the effect will simply be that of two single-clicks. You need to find out for yourself how rapidly you need to perform the clicks to achieve the correct effect.

Dragging means moving the mouse while an item is selected and the left-hand button is held down. This is a way of making the scroll bars on the directory or file displays move on the screen when the mouse is moved along your desk. The scroll bars are released when the mouse button is released. Dragging is also used on file moving and copying actions. When you drag any screen object, you position the cursor over the object, press down the left-hand mouse button and hold it, then move the mouse with the button held down. Do not release the button until you are sure that the object is where you want it.

- These three mouse actions are used also in many modern programs that do not make use of the Windows system. In particular you will see them used in word processors and in graphics programs, and in virtually all other programs that can make use of a mouse.

Running EASY-PC under Windows

If you use a modern computer that runs Windows 3.1 (or later) you will probably want to run EASY-PC from Windows rather than directly from MS-DOS. This can be done in two main ways, by using the MS-DOS item on the Main program group, or by setting up EASY-PC and EASYLASE as programs in the Application group. For occasional use, the MS-DOS group is easier.

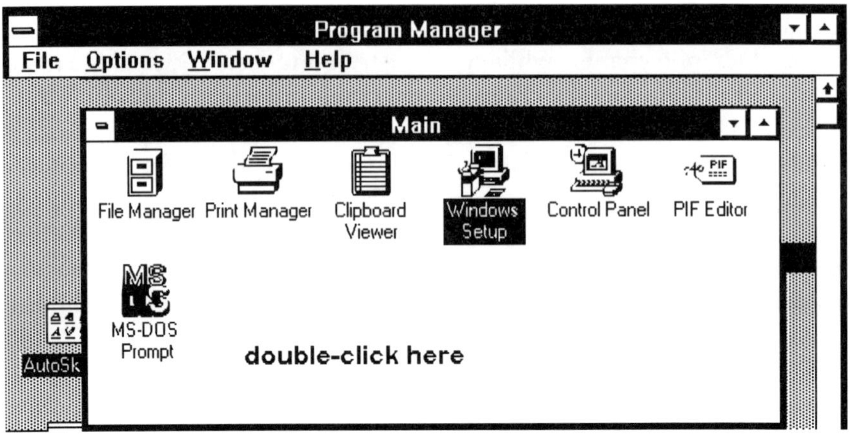

Figure 1.4 The Main group of Windows, which allows you to carry out an installation of EASY-PC under Windows control.

1 Bring up Program Manager and double-click on the Main group. Now double-click on MS-DOS Prompt (Figure 1.4).

2 This will return you to MS-DOS with the C:\WINDOWS directory in use.

3 Start EASY-PC by typing `C:\EASYPC\EASYPC` (press RETURN).

4 When you have finished using EASY-PC, and the MS-DOS prompt sign returns, return to Windows by typing `EXIT` and pressing RETURN.

The alternative is to place EASY-PC and EASYLASE (if you use a laser printer) into the Applications Group. This is normally done by using Windows Setup, but the EASY-PC programs cannot be installed in that way. Instead they have to be installed by using Program Manager.

1 With the Program Manager screen showing, open the Applications group by double-clicking on the name. Now from the menu select File and New. Click on Program Item, Figure 1.5, and then on OK.

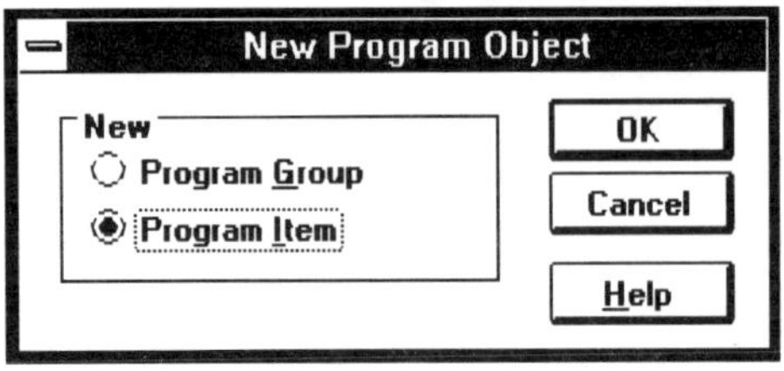

Figure 1.5 Clicking on the Program Item line of File New.

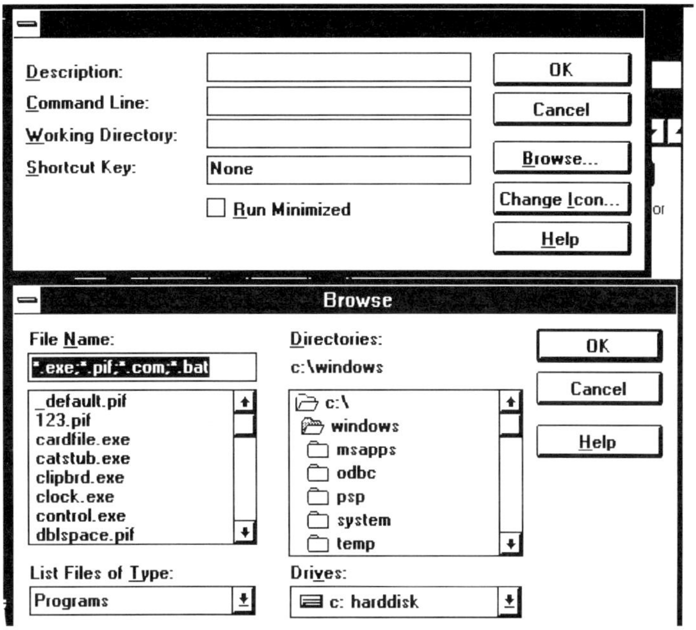

Figure 1.6 The Program Item form and the Browse display that allows you to locate directories and files.

2 You will see the form, Figure 1.6, appear. Unless you remember the EASYPC directory and filename that you have used, click on Browse to bring up the directory display.

3 Find the EASY-PC files in the Browse window, click on the EASYPC.EXE filename and then on OK, Figure 1.7. This will place the filename into the correct space in the form.

4 You can choose an icon by using Change Icon and Browse, Figure 1.8.

- Suitable icons and PIF files are currently available from Number One Systems Ltd.

5 You will see the icon added to the Applications Group, Figure 1.9. If you have not selected an icon, the icon will be that of the MSDOS box.

6 You must type in the name of the working directory, such as C:\EASY.

NOTE: If you are uncertain about using Windows as described, see the *Windows 3.0/3.1 Pocket Book* published by Butterworth-Heinemann. You can optimize the way that Windows will run EASY-PC (for example, by specifying background action) by using a PIF file, but this requires some knowledge and experience of Windows. The forthcoming Windows-95 will

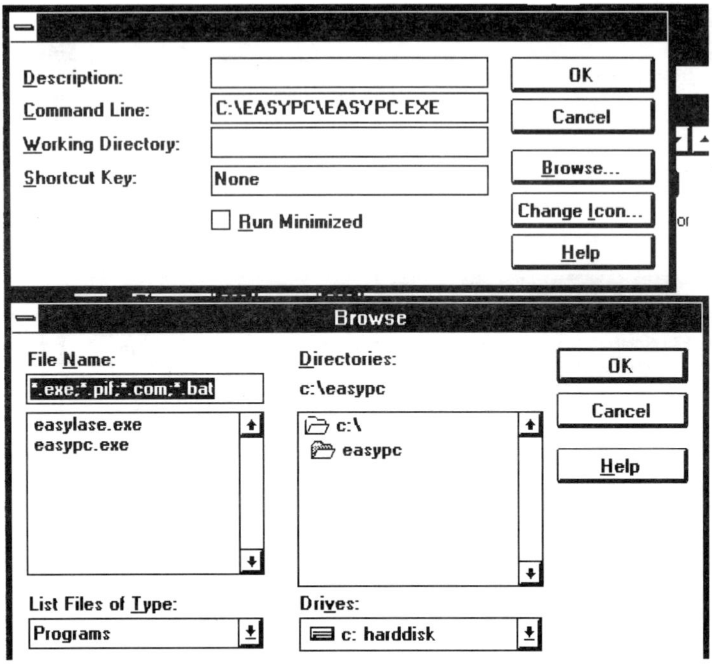

Figure 1.7 The entry obtained from using Browse. This avoids the need to type a name.

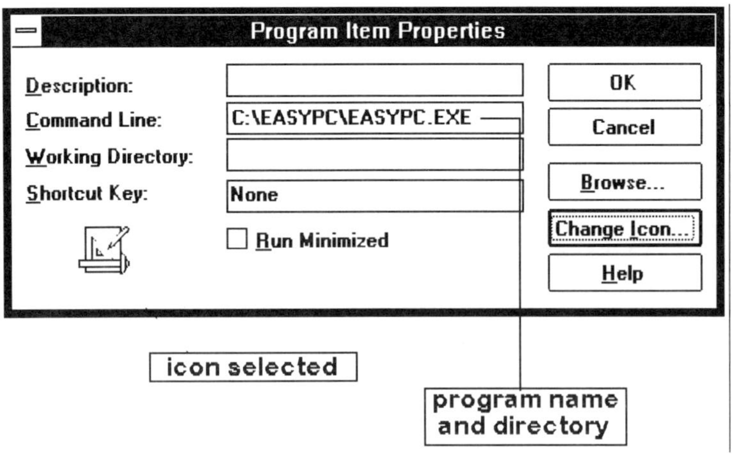

Figure 1.8 The icon selected from the Browse list.

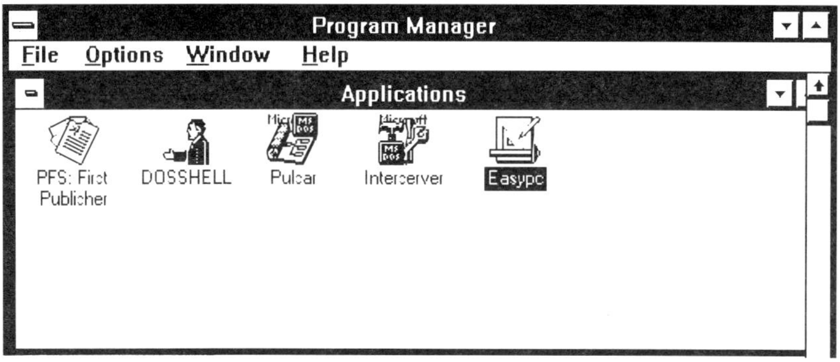

Figure 1.9 The icon in place in the Applications group. Double-clicking on this will run EASY-PC from Windows.

not require a PIF file and will make it much easier to run DOS programs like EASY-PC under Windows control.

Technical support

Technical support is available from Number One Systems for all registered users. If you find a problem which persists after re-reading the instructions and trying the action several times, contact the Customer Service team on (01480) 461778 or send a fax on (01480) 494042. When you contact Customer Service you should have the following information at hand:

1. Your Customer Reference Number (written into the manual).
2. The version numbers for EASY-PC and for your operating system (MS-DOS, PC-DOS or DR-DOS).
3. The contents of your CONFIG.SYS and AUTOEXEC.BAT files.

It is also very helpful when you are phoning in an enquiry if you can use a telephone that is within easy reach of the computer and the EASY-PC manual.

Upgrades and other offers are also available to registered users.

Virus infection

Virus infection is extremely rare if your computer is not connected to others (by a modem or a network) and if you load your programs only from disks that have been bought from reputable sources. If you do encounter a virus, confirmed by using anti-virus software, you should inform the Computer Crime Unit at New Scotland Yard. Even if the virus has done little damage to you, each piece of information sent to the Computer Crime Unit will help to track down another virus source. Viruses are costly to commercial users of computers, and the sharp shock of being caught is often a reasonable deterrent to the vanity of the writers of virus code.

A booklet called *Guidelines for Computer Professionals* is obtainable from:

The Computer Crimes Unit,
2 Richbell Place,
London WC1X 8SD

Tel: (0171) 230 1177

2 Getting started

Schematics

The EASY-PC manual deals very cursorily with schematics, but using EASY-PC for schematics is a useful way of gaining experience, and in many ways is a more useful start because you generally need to draw out circuit diagrams before you start designing PCB layouts. In addition, circuit diagrams are slightly simpler because they require just one print, there are no layers or silk-screens to worry about, and when you are learning to use any program a lack of complications is a definite advantage. In addition, when you upgrade to EASY-PC Professional you will normally draw the schematic first and use the facilities of EASY-PC Professional to prepare the PCB layout and also to analyse the circuit action.

- The screenshots that are used to illustrate this and other chapters in this book are in black and white for maximum clarity, and some use inverted shading (black and white reversed) because this avoids large areas of black that do not reproduce well. Where EASY-PC uses colour to distinguish items such as PCB tracks on different layers, screenshots in this book will use shading.

All schematic drawing starts, like other EASY-PC actions, from the main menu that appears when you start EASY-PC running (Figure 2.1). There are five items on this main menu, of which items 2 and 5 are the most important initially. Item 2 starts EASY-PC in schematic drawing mode and item 5 is used to end the program.

- At intervals, you will see a message at the foot of the screen:

  ```
  Save new version Y/N?
  ```

  which prompts you to take a backup file by using the Y key. At this stage, you can use the N key because you have not yet created a file to save. When you start to create diagrams, make use of this facility to ensure that you always have a backup file.

- You must never end EASY-PC simply by switching off the computer. Apart from anything else, you are liable to lose data if you do this.

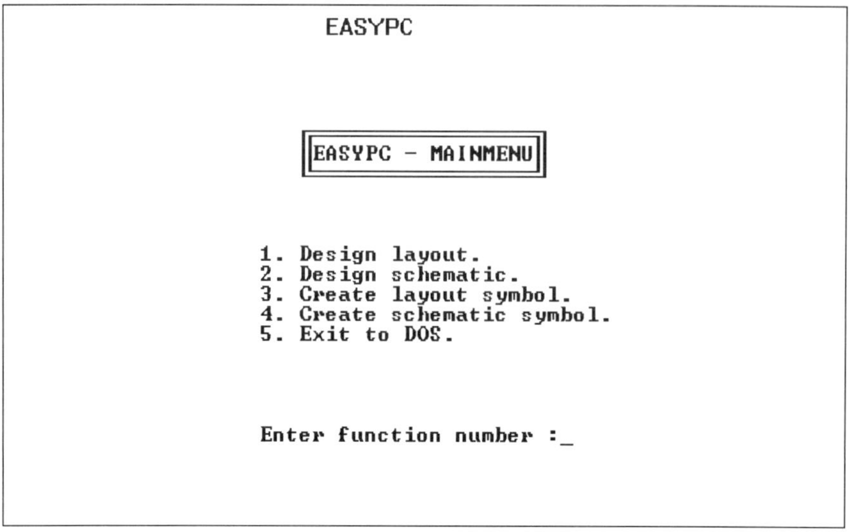

Figure 2.1 The main menu of EASY-PC as you see it on screen.

- If you find that the EASY-PC display in colour is not clear enough, you can alter the colours by pressing the SHIFT-F key (this will be referred to as ↑F from now on).

Items are selected by typing the number – you cannot click with the mouse as you might have been accustomed to in other programs. When you type 2 the screen changes dramatically, and if you have been used to the Windows type of screen with a menu along the top you might wonder where everything is (Figure 2.2).

- The screen at this stage is mainly black, with the three small squares and the rectangle in a light colour.

The menus are obtained by clicking on the small squares, and the rectangle in the centre of the screen is your drawing space. If this looks too small, be assured that you can magnify this view on the screen when you need to. What is difficult to adjust to is the fact that there is no initial option for setting a board or drawing size – this is something that you do when the drawing or board layout is printed, not before. The drawing space represents a board of 17" x 17", and this is fixed.

- If the schematic, when printed, would look too small, you can alter the scale of printing. This would not be desirable for a PCB because components such as ICs have fixed sizes, and the printing must be done to the 1:1 scale so that the dimensions of drilling-holes for the

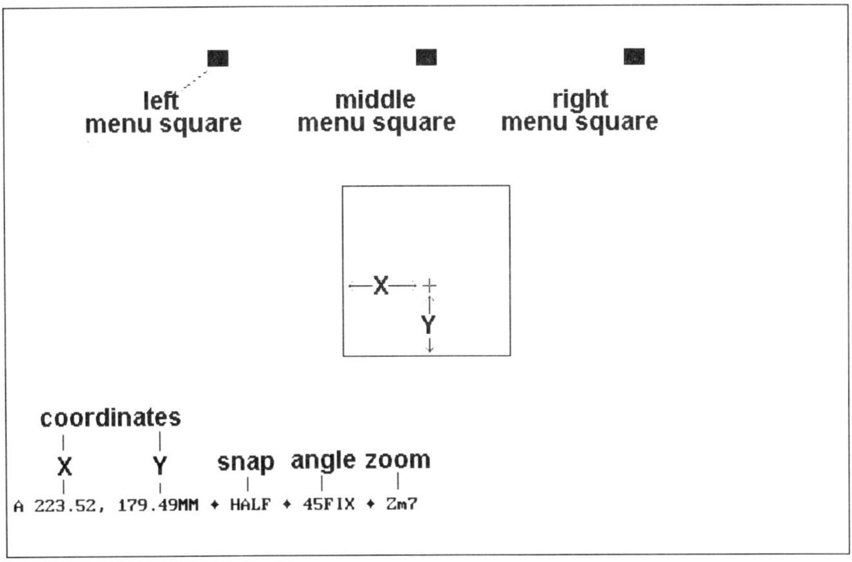

Figure 2.2 The screen that appears when you select Schematic as the drawing option.

components are correct. Scaling is appropriate for schematics, however.

Zoom, Grid and Scales

When you place the mouse cursor over the right-hand box at the top of the screen, a menu will appear. This menu can be used by clicking with the mouse, by moving the cursor with the cursor keys and pressing RETURN or by pressing letter keys (with or without SHIFT). The use of the letter keys is by far the more convenient, and the menu (Figure 2.3), contains a reminder of the keys that you can use. There is an alphabetical list of actions with their keys in Chapter 6.

- The menu disappears if you press the ESC key or click with the right-hand button of the mouse.

Copy the following list and keep it close to the keyboard. When you start to use EASY-PC you will find it simpler to make use of the menus that appear when the cursor is placed on one of the square blocks (menu boxes). Note that the central box has no effect unless you are editing, see later. As you become more familiar with EASY-PC you will find it convenient to remember some key actions, which are quicker. The EASY-PC manual provides a keystrip reminder for the main keys (the F-keys), and this strip should be attached to your keyboard.

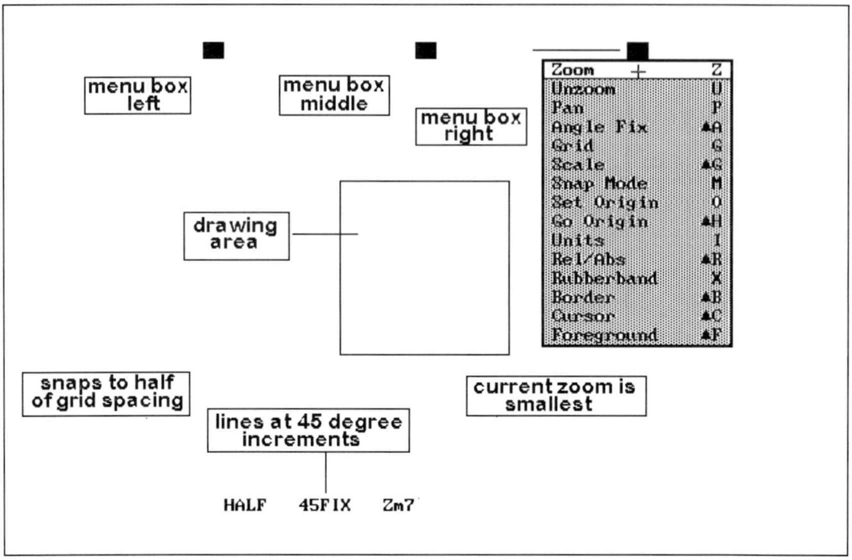

Figure 2.3 The right-hand menu expanded, and a guided tour of the screen. Note that nothing appears in the centre menu unless an editing or filing command has been selected.

The list consists of:

Zoom	Z
Unzoom	U
Pan	P
Angle Fix	↑A
Grid	G
Scale	↑G
Snap Mode	M
Set Origin	O
Units	I
Relative/Absolute	↑R
Rubberband	X
Border	↑B
Cursor	↑C
Foreground	↑F

Note that the ↑ means that the SHIFT key is also held down along with the letter key.

You may find when you browse over these commands that some appear to have no effect. That's because commands such as Grid are used only on more magnified views; and some other commands enable a menu to be activated by the middle box. For the moment we want to look only at the Zoom, Grid and Scales commands.

The *Zoom* command, as the name suggests, alters the magnification of the screen view – the 17" x 17" maximum area is not altered, only your view of it. The command can be executed in several ways, but the most useful are the use of the Z key and the use of the number keys from 1 to 7. The smallest level of magnification is Zoom 7, which is what you see by default when you start up EASY-PC. You can change to any other level by pressing the number key corresponding to the zoom level you want (1 to 7), or you can use the Z key to change the zoom level by 1. Pressing Z once will change from 7 to 6, press again to change from 6 to 5 and so on. If you want to change zoom level by one unit in the other direction, use the U (Un-zoom) key.

- These (and other) changes take place only if the cursor is within the drawing area, and the position of the cursor determines the position of the centre of the zoom.

Figure 2.4 shows the effect of zooming up to level 6 and pressing the G key to bring up the grid. The grid is a guide but it also shows snap positions. A snap position is where the cursor will move to when you move the mouse or use the cursor keys, and the point of using snaps is that the cursor cannot be moved to any intermediate positions, even if you think that you are in an intermediate position from the screen view.

The default of EASY-PC is to have the snap positions at half of the grid interval, which is the reason for the *HALF* message at the foot of the screen.

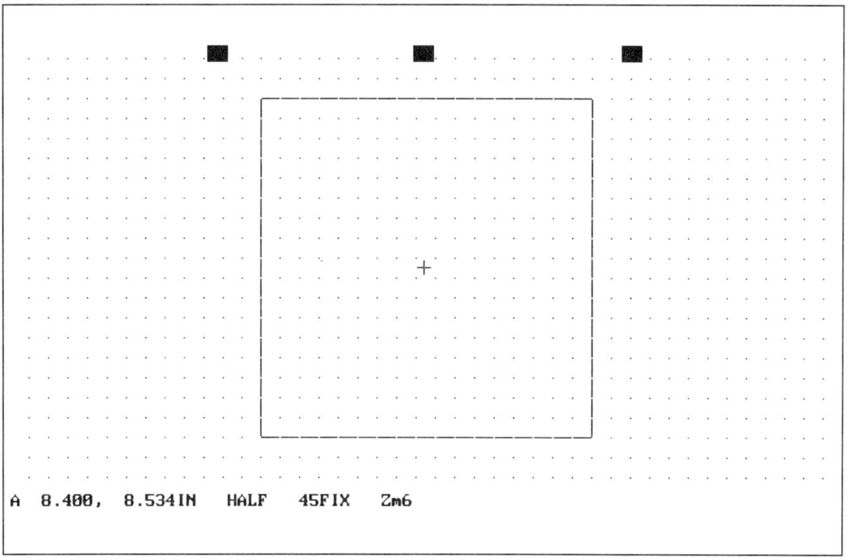

Figure 2.4 The drawing area expanded to zoom level 6, showing the grid that appears when the G key is pressed at this zoom level.

The other possible messages at this point are *GRID* (snap to each grid point), *QTR* (snap to quarter of grid distance), *FREE* (no snaps) and *NEAR* (used for PCB layout only, goes to the nearest node point from where the mouse button has been clicked while laying a PCB track).

- The grid that you see on the screen, however, is not as fine as the grid that is used for snaps. At zoom level 1 it is, but because the grid is set to an interval of 0.1" and would be impossible to show precisely on the screen at zoom levels 5 or 6, what you see at these magnifications is a set of dots at 1" intervals, only every tenth dot of the real grid. This can be confusing if you are not aware of it.
- The Scales command (use the ↑G key) will replace the grid by a set of scale markings around the screen. This option is not so generally applicable to schematics, but it can sometimes be useful in conjunction with the large cursor: see later.

You will find it useful to change the Angle Fix for schematics. The Angle Fix means the angles that a line is permitted to take when it is drawn from another line, and the default is 45°. If you change this to 90° you will find this more useful for schematics, because it is very seldom that we want to use 45° lines – the main exception is for a discrete component multivibrator. To change the angle fix use the ↑A key. The options are 45FIX, 90FIX and MULTI and you will see EASY-PC cycle through these options as you press the ↑A key. The MULTI option allows you to draw lines at any angle, but remember that if you are snapping to grid points some angles cannot be used. In any case, the use of MULTI is likely to produce some lines that look very ragged on screen, though the printed quality is much better.

- If you have been clicking with the mouse and you now find that lines have appeared, clear the drawing space by using the ↑F3 key.

Symbols

Schematic drawing with EASY-PC consists of placing component symbols on the grid and then drawing lines between them. Figure 2.5 shows the menu for the left-hand square, which shows the F8 key as a shorthand for *New Symbol*. Note that this view uses the largest magnification, zoom level 1. The order of events is that you put the cursor where you want the symbol to appear, press the F8 key and type the symbol name.

- This system is used throughout EASY-PC and is by far the easiest way of working, because it avoids the need for too much wrist action with

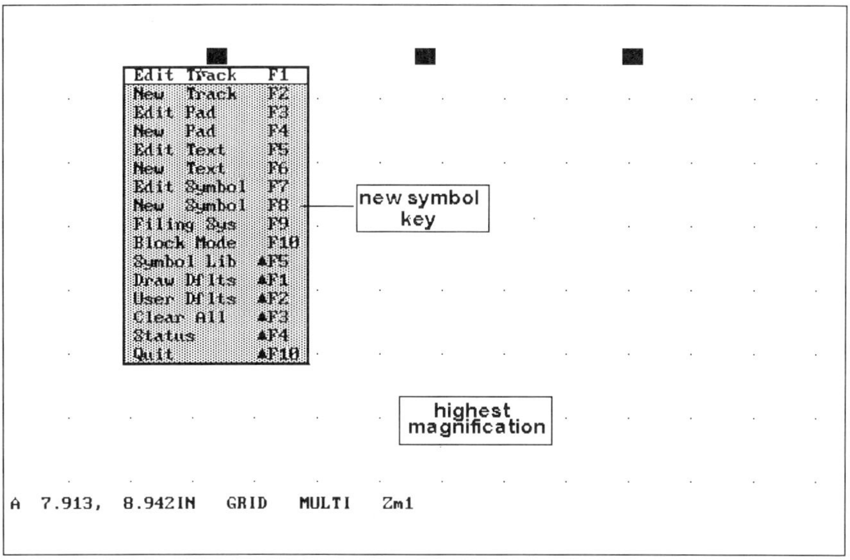

Figure 2.5 The menu that appears when the cursor enters the left-hand menu box. This is mainly concerned with editing, and the use of the F-keys is usually more convenient.

the mouse. The keystrip provided with the manual shows the F-key actions.

Remember that when you use the F8 key to obtain a symbol you need to specify the filename.

- EASY-PC will, however, supply the correct extension letters of SLB for a drawing (schematic) symbol. This adjustment is made automatically by your menu choice – when you switch over the PCB layout, the extension letters will be LLB.
- Note that the F8 key supplies a New Symbol. If you want another of the same type you can use a copy (Repeat) command rather than using F8 to load in another symbol of the same shape.

The symbol will appear at the cursor position – meaning that some part of the symbol will be snapped to the cursor position. For a simple example of an NPN transistor, it is the end of the base lead that is located at the cursor position, and every symbol has a corresponding location point (or origin) like this – you have to learn where they are by experience.

The actual position, however, is not important, because when you have loaded in a symbol (before you press any other keys) you can click at any point on the grid (or wherever you can snap to) and the symbol will move

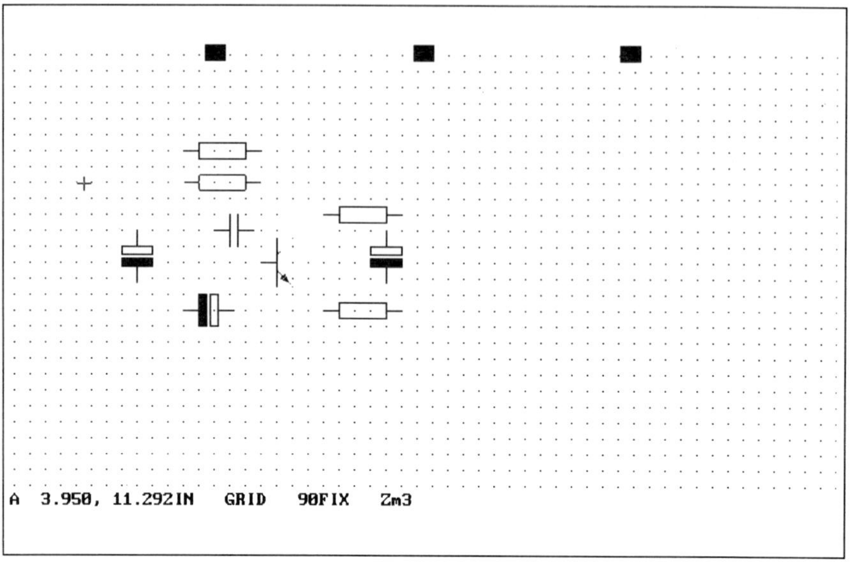

Figure 2.6 A drawing started by placing components in the drawing space.

there. Figure 2.6 shows an NPN symbol placed on a grid point in this way, with some other components.

- The symbol can be moved as much as you like by moving the cursor and clicking, until it is locked into place (anchored) by pressing the right-hand mouse button or the ESC key.

Looking at the symbols in the example, the resistor symbol has been copied several times. This is done as follows, and the example is a useful one because it will help you to understand the ways that EASY-PC handles its commands:

1 Place the cursor over the resistor.

2 Press F7, the Edit Symbol command.

3 Press the R key, for Repeat.

4 Move the cursor to where you want the new copy to appear and click the left-hand mouse button (or press the RETURN key).

5 Anchor the symbol by pressing the right-hand mouse button (or pressing the ESC key). Alternatively, if you want another copy of the symbol, repeat from step 3 above.

If you do not move the cursor far enough and as a result you find that one copy of the resistor is over the other when you press the left-hand mouse button, simply click at a point farther away and the new copy will move there. This will work until you press the right-hand mouse button to anchor the symbol. Practice this copying action, and note that you return to the New Symbol action automatically if you click at a point without pressing the F7 key (or other F-key) first.

- Use F7 followed by ↑Z (Zap) to delete the component that the cursor is over. You will be prompted (Y or N) to confirm that you really want to do this.

Remember also that you can use the ESC key or the right-hand mouse button to escape from a command. For example, if you have earlier used the F8 key to insert a new component and you later click at a point and see the request for a component library filename, you can escape from this – you will still be in component insert mode, however, until you use another key. The F7 key temporarily changes to edit mode, and other keys permanently escape into other actions.

The next step with the example is to turn the resistors and the electrolytic capacitor the right way round. When you place the cursor on a component and use the F7 key, this activates the middle menu box (which is not used until you are editing). The list of key options in this box for symbol editing is:

Edit Text	T
Current	C
Layer	L
Angle	A
Flip	F
Repeat	R
Zap	↑Z

The Layer option is not applicable for a schematic.

The Angle command (cursor over component, press F7, then A) allows you to enter an angle as a number in the range 0 to 3. Each increment represents an angle of 90° anticlockwise from the position in which the symbol is normally inserted and, for our purposes, the number 1 is needed to change a horizontal resistor into a vertical resistor.

Figure 2.7 shows the components rotated as required and shifted into more appropriate positions. Remember that you move a component by placing the cursor over it, pressing the F7 key and then clicking where you want to component to be put – you can click again (without using F7) if you want to alter the position. When a resistor is turned vertical, its clicking location (origin) is at the lower end of the connector. The electrolytic capacitor symbol (filename CP) needs a rotation angle of 1.

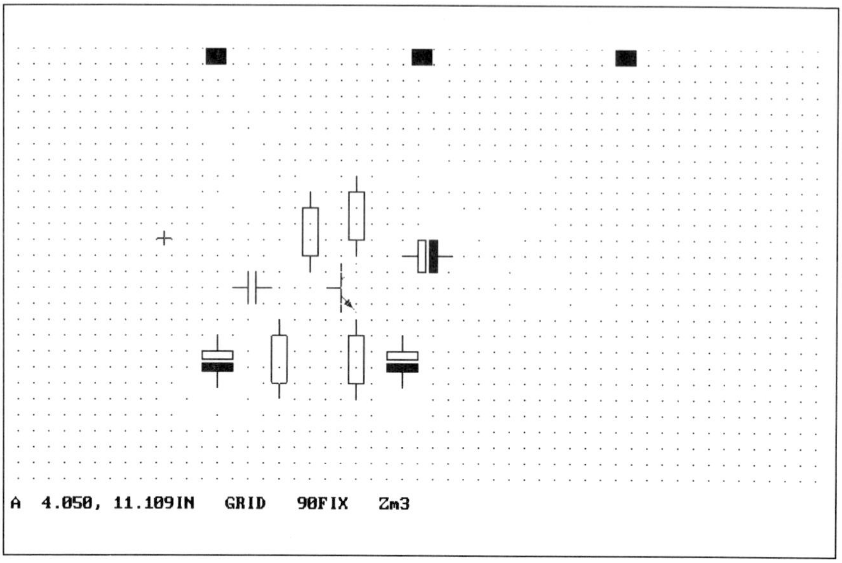

Figure 2.7 Rotating components to fit into the correct places in the diagram.

- If you have already rotated a component and want to rotate it again, remember that the angle number is absolute, not relative. For example, if you earlier applied a rotation of 1, you need to use the number 2 if you want to turn the component through another 90°.

Interconnections

When the components are assembled into position, making use of snap to ensure that connections will line up, you can make the interconnections. This is done using the F2 key, labelled as New Track. Before we can start on this, however, we need to take a close look at the diagram as it currently exists.

The components shown in Figure 2.7 have bee placed using a snap setting of *GRID*. This places the location points or origins of components on to grid points, but it does not ensure that other connection points on a component will also be on grid points. As it happens, they will be for some components, but the NPN transistor with its base lead on a grid point has the ends of its emitter and its collector leads on points that fall between grid points. You must therefore switch the snap mode to *HALF* before you try to make interconnections, otherwise the connections will not join the emitter or collector leads.

- You can change the colour scheme that EASY-PC uses by pressing the ↑F key until you find a set that you want to use.

Making interconnections is baffling until you grasp the rules. In particular, you find that when you click the mouse button, you create a line from wherever the cursor was to where it is now – and that line is not always wanted.

- Where you click the mouse button (or RETURN key), a *node* is created. This (invisible) node point contains all the information on a line, such as width, colour etc. This point is very important, because when you delete a node (see later), its information is also deleted. Similarly when you copy a node, its information is also copied.
- You can delete such a line when the cursor is on a node by using F1 and then the letter D to delete the node, followed by ↑Z to delete the line. You will be asked to confirm the ↑Z action.
- You need not locate the cursor directly over a node, because EASY-PC allows you to specify a minimum pick-up distance, typically set at 1" (it can be varied from 1" to 4"). If the cursor is within this distance from a node then the effect will be as if the cursor were directly over the node.
- Making interconnections is always easier if you switch on rubber-banding with the X key. This allows you to see the line that will be created before you fix it into place.
- A line is drawn from one point, a node, to another node. The first point is the *home node* for that line and the last point is the *end node* for that line. The Home (H) and End (E) commands will place the cursor on the home and end nodes, respectively, for a line. From the current cursor point, you can also go one node back with the Back (B) or one node forward with the Next (N) command. If you have moved the cursor with the mouse or the cursor keys, you can locate it on its correct position in a line by using the Current (C) command. You can use these commands when you are drawing a line or while you are editing a line.

1 Start your interconnections in this example by pressing the F2 key and then the ESC key. This latter key will ensure that no line is drawn from the present cursor position.

  - Do not use the right-hand mouse button in place of the ESC key because if you press the left-hand mouse button by mistake you will create a line of zero length.

2 Place the cursor over a point where an interconnecting line will start. Click the left mouse button.

3 Place the cursor over where the line is to end and click the mouse button again. The line will appear. If you are using rubber-banding it will have been visible (as a thin white line) while you were moving the cursor.

4 Click the right mouse button or tap the ESC key to make sure that a new line will not start from the last point visited.

- The use of the ESC key, or the right-hand mouse button, closes the editing actions for the line. If you want to make further changes you must place the cursor on a line node and start line editing (F1 key) again.

Figure 2.8 shows the lines of the example drawn in.

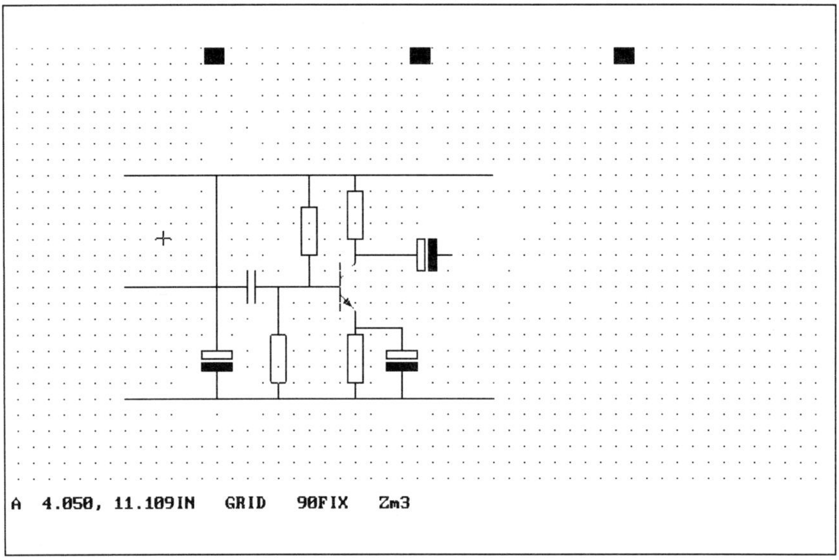

Figure 2.8 The component ends joined by lines, using the New Track option of EASY-PC.

Now for some refinements. We need to show a ground (earth) symbol, and this can be added by using the F8 key for new symbol and specifying the name 0V (zero volts) for the older earth symbol, or *EARTH* for the later symbol (chassis earth). Both of these symbols are shown in Figure 2.9.

In addition, many schematics use a heavy dot to indicate a circuit T-junction and this convention is often insisted on when diagrams are to be reproduced in books or magazines, because it removes any ambiguity about whether lines cross or join. You can use a PCB pad symbol as a dot, because these PCB symbols are still available even when you are using EASY-PC for drawing a schematic.

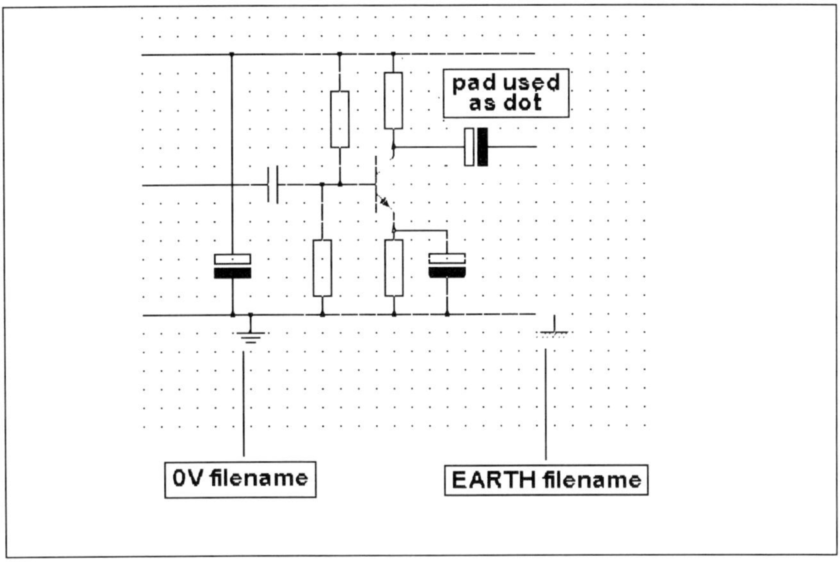

Figure 2.9 The two earth symbols and the use of a pad as a dot.

The pads are used as follows:

1 Place the cursor away from the drawing. Press the F4 key for New Pad. This will put a standard pad at the cursor position. This pad is, however, too large, and a centre hole is not wanted.

2 With the cursor over the pad, press the F3 (Edit Pad) key. Now press the W (Width) key.

3 In response to the `Please key in width` message at the foot of the screen, reply with 5, which is a reasonable size to make a dot in a schematic.

- This figure of 5 assumes that you are using the default set of pad sizes that EASY-PC installs. If you have altered the defaults (see later) you will need to choose whatever number corresponds to the size you need.

4 Now bring down the centre menu and select Variant. You will see a selection of pad shapes, Figure 2.10, from which you can use A or C, since these (and several others) have no hole.

- You do not see the pads without holes when you use the V key in place of using the menu.

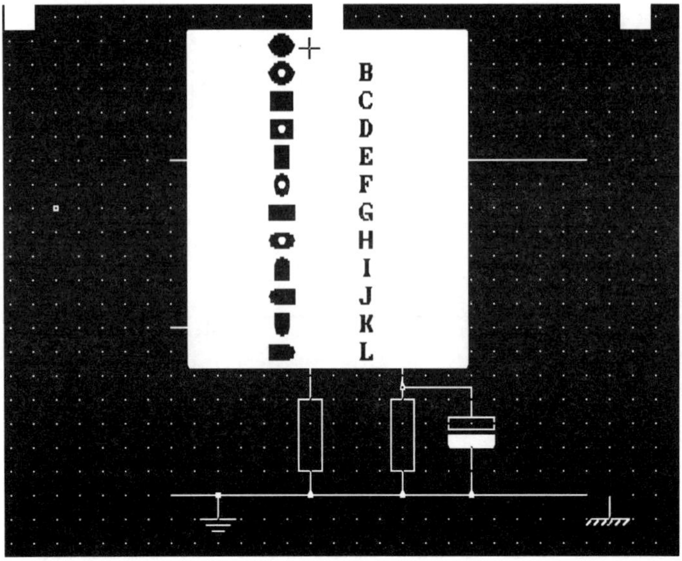

Figure 2.10 The selection of pad shapes.

5 When you have the correct size and type of pad, place the cursor over the pad, and press the R key (to Repeat). Put the cursor where you want a dot placed and click the left-hand mouse button.

6 Repeat the use of the R key and the mouse button until each dot has been placed. Use the ESC key or the right-hand mouse button to escape from the action. You will be ready to load another symbol until you use another F-key command.

- Remember now to delete the first pad you laid down away from the drawing space.
- Remember that if you need any other Edit Pad action, such as Delete, you need to press the F3 *Pad key before pressing the D key.*
- If you try to use EASY-PC without a mouse you will have to use the cursor (arrowed) keys to move the cursor and the RETURN key to make a command execute. If you are using a fast machine you will find that the cursor keys are very sensitive, and you have to tap them very briefly if you want to move only a short distance. You will be thankful for snap, but even more thankful if you have a mouse.

Using Zoom and Pan

When you are working with fine detail, such as using pads as dots, you will find it easier if you use a zoom to magnify the image. Figure 2.11 shows the amplifier circuit at zoom level 2 – the extremes are hidden, but all the main connections are visible. You can shift the view, as distinct from altering the zoom level, by using the Pan command. This can be obtained either from the right-hand square menu or by using the P key.

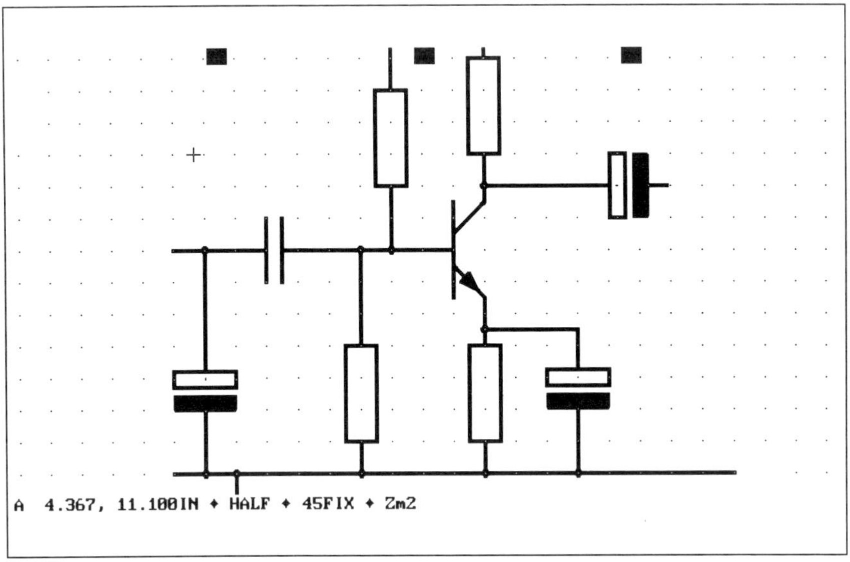

Figure 2.11 Zooming to level 2 to see the schematic in detail. Remember that you can use the Pan command to shift the centre of view.

Place the cursor to the right of your drawing, and press the P key to shift the drawing left. Place the cursor to the left of your drawing, and press the P key to shift the drawing right. What this amounts to is that pressing the P key makes the cursor position become the centre of the view. By using both zoom and pan, you can focus on any part of a large diagram and carry out detailed insertion or editing actions.

Rubber-banding in detail

The normal method of connecting components works, as described earlier, by clicking in turn on the points that are to be joined, and the snap system ensures exact location. An alternative way of viewing the joining of components is to use rubber-banding, switched on by pressing the X key (it will remain on until you cancel by using the X key again).

When rubber-banding is switched on, clicking on a point will start a (white) track line which will then follow the cursor and become fixed when you click again, Figure 2.12. This lets you see what is going on before you click to fix the line, and you will see at once, for example, if you have forgotten to set the snap points correctly. It also reveals what angle mode you are using – if you have set for 45°, for example, you will see sections of line at this angle as you move the cursor.

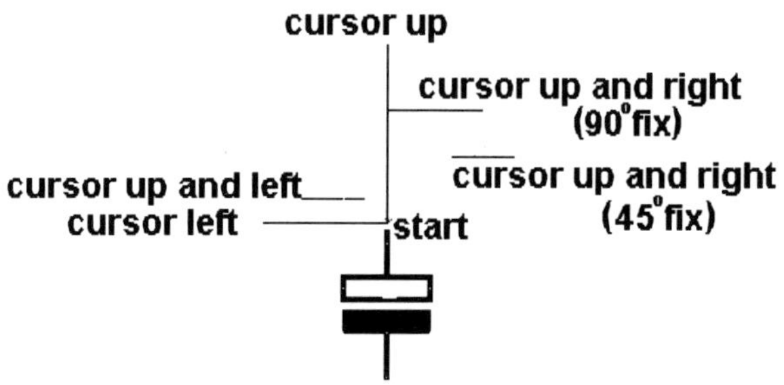

Figure 2.12 Using rubber-banding. This allows you to see what course a line will take before you fix it in place.

Rubber-banding is more emphasized for its use in PCB layout but it can be very useful for schematics, particularly if you have a crowded layout with many lines crossing. Using rubber-banding allows you to see what the track will look like before you fix it into place, and because the line is white it is easy to see on a dark background. The line will change to red (or whatever colour you are using for such lines) when you click to fix it. When you have rubber-banded one line, another line can be dragged out from the last point where you clicked, unless you use the ESC key or the mouse right-hand button.

- Remember that you can press the X key to switch rubber-banding off again.
- A very useful action that you can use when you are laying down track with rubber-banding on is Track Flip. When you have a track that contains one straight section followed by another at a 45° angle, using

the Flip command (key F) will reverse the arrangement, reversing the positions of the angled part and the original straight part, Figure 2.13. The Flip command does not work in Track Editing.

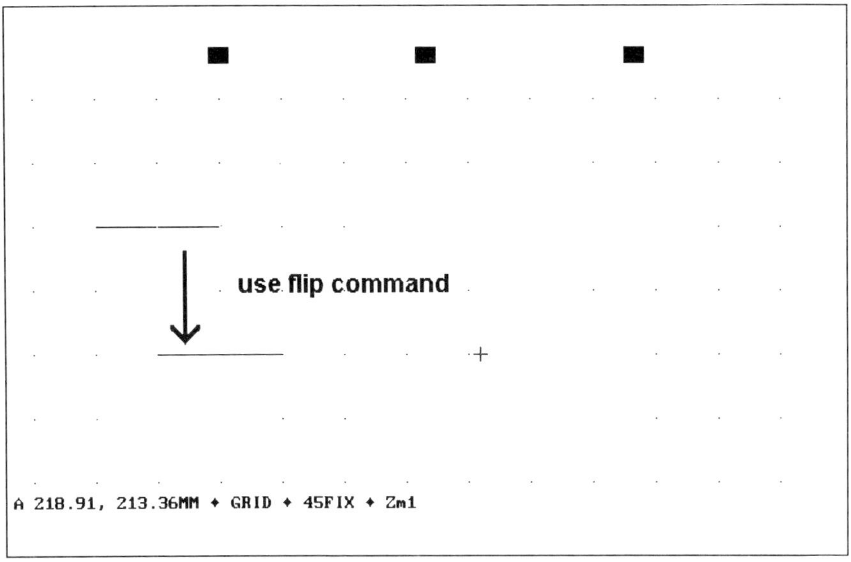

Figure 2.13 Using Track Flip to reverse the way that a track is taken from one point to another by way of a 45° change of direction. This works only when you are laying down track lines, not when you are editing.

Scales

The use of scales is an alternative to the grid. Scales are markings, at fixed intervals of 0.1" (2.54 mm), placed around the drawing area (Figure 2.14), which are obtained by pressing the ↑G key. The scale will always replace the grid, and you can return to the grid by using the G key.

Scales are used more for PCB layout than for schematics, but they can be useful as a way of guiding you when a layout is unusually dense or if the presence of the grid dots makes it difficult to see some details, such as the dots in T-junctions.

Saving work

The culmination of all this work must be a file saved on the hard disk (or a floppy), otherwise all of your work is pointless. We'll look at printing in Chapter 3, but even if you think that a schematic is a one-off, you should

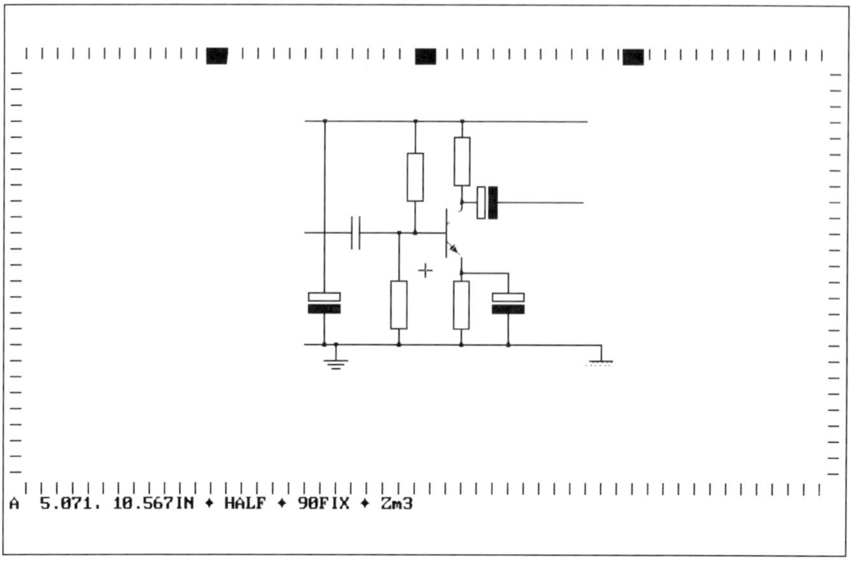

Figure 2.14 Using scales in place of the grid as a guide to alignment.

not simply print it and then forget it. You might, for example, find that the printed diagram did not look as you expected it and that you need to make some changes. This is not exactly easy to do if you have not saved the work as a file. The other point is that there can never be a diagram that is not re-usable, because you can edit symbols and connections to convert one diagram into another.

Saving a diagram is done by using the Artwork Filing System, obtained by pressing the F9 key. This allows the central box object to reveal a menu when the cursor is placed over it, Figure 2.15.

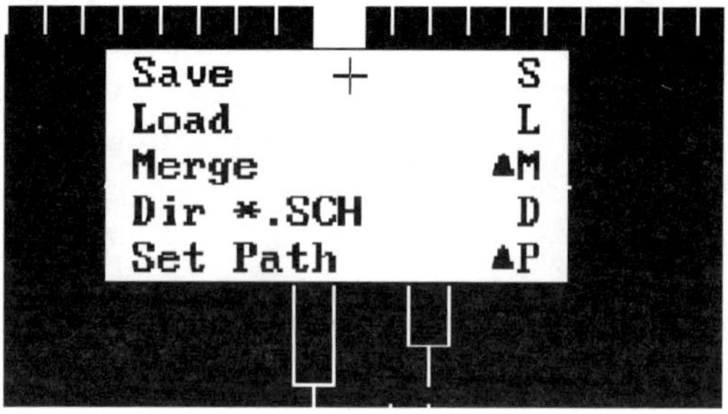

Figure 2.15 The menu that appears for the central menu box when the F9 key has been pressed for Filing System.

Save is used to save your file, and when you select this from the menu, or press the S key after the F9 key, you will be asked for a filename to use. If the file has been saved before, you will see its filename appear but if this is a new file you may see a name that has been used before. If you have not saved or loaded any drawing files in this session you will see the default name NEWBOARD.SCH. To change this, simply type the name you want to use, omitting the SCH extension letters, and then use the DELETE key to remove the default name of NEWBOARD.SCH. EASY-PC will automatically add the SCH extension to the filename when you press RETURN.

If a file of the same name already exists you will be given a chance to try again or to press the RETURN key to write over the old file (whose extension will be changed to BK2). This allows you to recover the old file by renaming it. Only the most recently replaced file is preserved in this way.

- You should also make a backup of files on a floppy disk. You might not want to do this each time you save a file, but you should always back up files that you have created before shutting down EASY-PC for the day.
- EASY-PC automatically makes a backup on the hard drive when you alter a file and save it again, and these automatic backups (see later) allow you to return to an earlier version if you have altered a file and want to undo your alterations.

A file that you have saved in this way can be loaded in again by using the F9 L key sequence or by using the *Load* menu command following F9 and moving the cursor to the File menu. As before, you need to type the filename and press the RETURN key to load the file.

- You should keep a note of filenames and what circuits they contain. EASY-PC Professional allows you to browse over a list of filenames and select by clicking.

The DIR *.SCH (key D following F9) menu action will allow you to see a list of the drawing files that you have. This appears like an MS-DOS DIR listing, Figure 2.16. You can print out this screen in the usual way (put the printer on-line and press the Print Screen key), or take a note of the names that are listed.

The *Set Path* action (use ↑P following F9) allows you to make use of a directory other than C:\EASYPC for your drawing files. You can edit the default C:\EASYPC to show a path to another directory, but the path name does not follow the usual MS-DOS convention exactly.

- A name containing a path is acceptable in a form such as C:EASYPC\DRAWINGS but the more familiar format of C:\EASYPC\DRAWINGS is not.

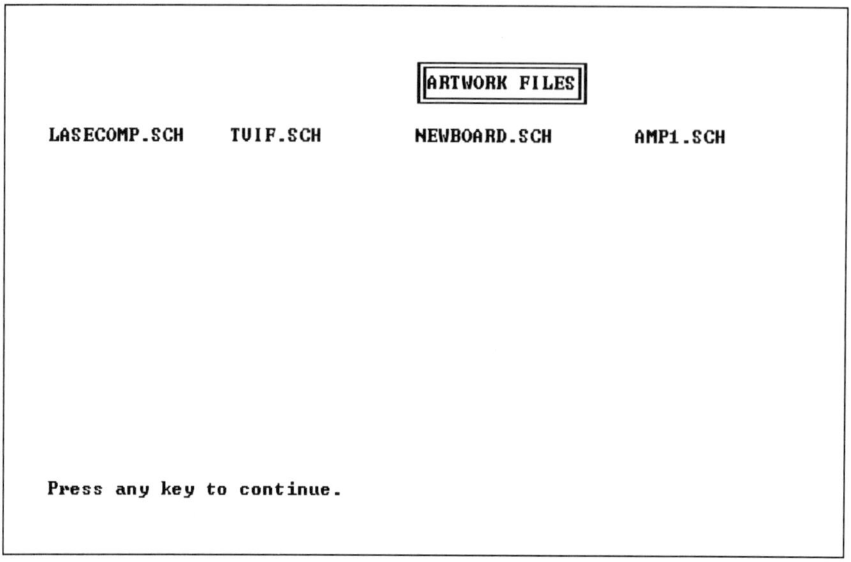

Figure 2.16 The DIR *.SCH listing, which corresponds to an MS-DOS DIR listing for the schematic files. EASY-PC Professional allows you to load a file by clicking on its name.

Finally, the last option in the File set is *Merge*. When you use a Load, the new file replaces the one that you have been working on, but Merge allows a new file to be added to the existing one. The file is always merged in at the location it had when it was saved.

- Merge is particularly valuable when you have a portion of schematic which is used on all of your drawings, such as a power supply.

Merge will always place the merged file into the drawing area at the position where the file that is being merged was placed when it was saved. You will find merging much easier if you adopt a consistent positioning of work. For example, if you always start a drawing at the bottom left-hand corner of the drawing space, use the top right-hand corner for merging. When you have a portion of drawing that you want to merge, move it to this (top right-hand) position and save it, so that it will be well clear of your main working area (bottom left-hand) when you merge it into another drawing.

3 Enhancing schematics

Adding text

Text is a very important part of a schematic, if only to label components. You will want, at least, to label components with references such as C1, R2 and so on, and on the smaller diagrams you may be able to make space to insert component sizes along with the circuit references. On more tightly-packed schematics, it is more usual to put the references only on the circuit itself, and use a separate table of component parts. This table can be placed immediately under the drawing or to one side, or it can be quite separately printed, depending on the space you have available.

- When you are using a text menu, the T symbol will appear as a separator between the information items on the bottom (status) line of the screen. Letters L, P, T and S identify Track Line, Pad, Text or Symbol when you are editing. A red diamond shows that no item is selected.

Placing text on to a diagram starts with placing the cursor at the point where you want text to appear. You do not have to be too precise about this, because the text can be shifted later, but the closer you can get to the final position the better, because it saves time.

Figure 3.1 shows this action under way. The reference R1 has already been put in, and the size of the text is, by default, right for the size of the components that are held in the symbol library. Another reference is being put in, with the cursor placed next to a resistor symbol, the F9 key has been pressed and the reference R2 has been typed (bottom line). Pressing the RETURN key will place this at the cursor position.

When you have placed a piece of text in this way, it will follow the cursor, and move to wherever you click the mouse button (or press the RETURN key). You then fix the text in place by pressing the ESC key (or right-hand mouse button). The appearance of text may look ragged until you fix it in this way. When you want to place another piece of text you must position the cursor (without clicking) and click the left-hand mouse button (or press the RETURN key) to get to the text editing line at the bottom of the screen. Figure 3.2 shows the whole diagram lettered in this way.

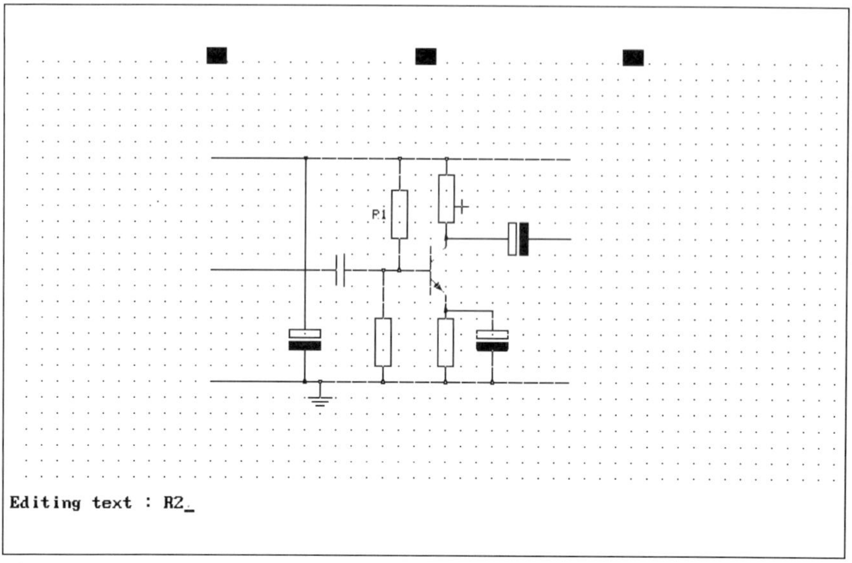

Figure 3.1 Starting the entry of text. The default size of the text is ideally suited to labelling a schematic or a PCB layout.

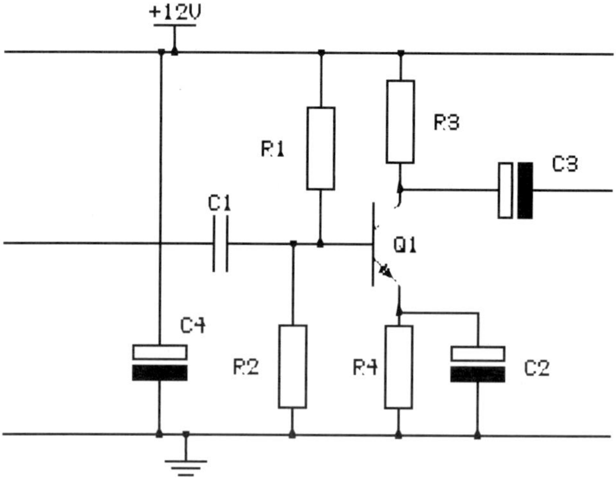

Figure 3.2 The diagram completely lettered with component references. On a simple diagram, you could also put in component values.

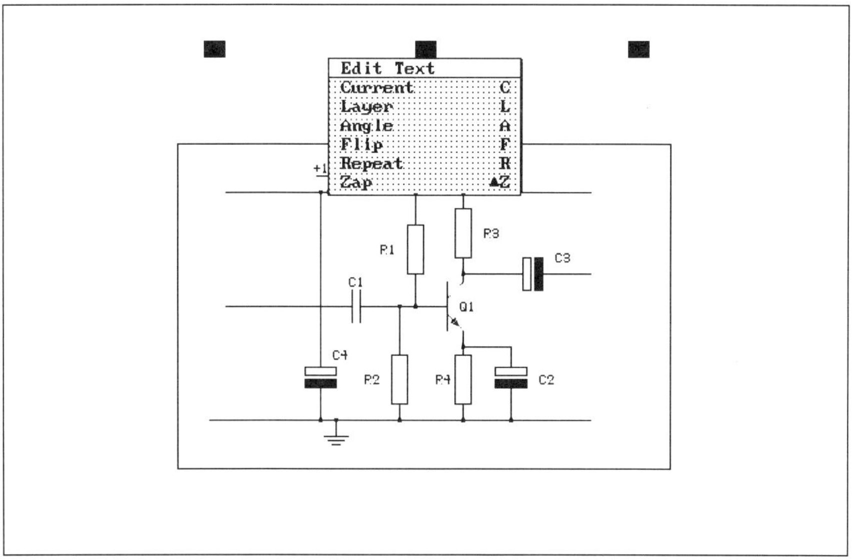

Figure 3.3 The text editing menu options that appear in the central menu when the F5 key has been used.

You will normally want to use the default text letters, but when you use the F5 (Edit Text) key you have a new menu of text effects available, Figure 3.3. The one that you are most likely to use at this stage, following F5, is *Edit Text*, obtained with the T key. This allows you to alter the text wording (as distinct from position, size etc.)

- All of these actions are also available after you have used the F6 key and until you press the ESC key (or right-hand mouse button).
- When you want to alter the position of text, remember that the origin point for a piece of text is the left-hand end of the text. If you just put the cursor over any part of the text it might be too far from the origin to pick up the text in a move action.

The procedure is to place the cursor on the text that you want to change. Then use the F5 key followed by T for Edit Text, and type in the new text, using the DEL key to delete the older text in the editing line at the bottom of the screen. The text is replaced when you press the RETURN key.

- Note that when you change text you may see the old text remaining. If this is a problem, use a pan action. Place the cursor in the centre of the screen and press the P key. This is generally applicable for refreshing the screen display after any editing. Another option is to use the zoom number key for the zoom level you are using. For example, if you are using zoom level 3, press the 3 key.

The C key (the Current menu item) will make the cursor move to the starting point of the last piece of text you worked on. If you have made a copy of this text using the R key, the copy will become the current text, and will be the text that the cursor is placed on when you use the C key.

The L key (Layer) has an effect on text colour, though layers are not really applicable to schematics. You can enter the numbers 0 or 1 in response to the question about layer when you press the L key. When you use the default of 1, any text that you enter will be in red. When you use layer 0, the text will be in white.

- You can also change the colour of existing text by changing its layer in this way. The use of layers is intended more for PCB layout, with layer numbers 0 to 8 available. For PCB use, you can place text on a copper layer (so that it is etched in copper). This is useful for identification text, such as ID numbers, that you want to have in a more permanent form, since, unlike a silk-screen image, it cannot be rubbed off.

The *Size* (key S) command allows you to use a size number from 1 to 4. The default is 1, which is the size best suited for the size of symbols, but if you need to make tables of components with larger print in the headings, you have sizes 2 to 4 available. Figure 3.4 shows these size options with a resistor symbol for comparison.

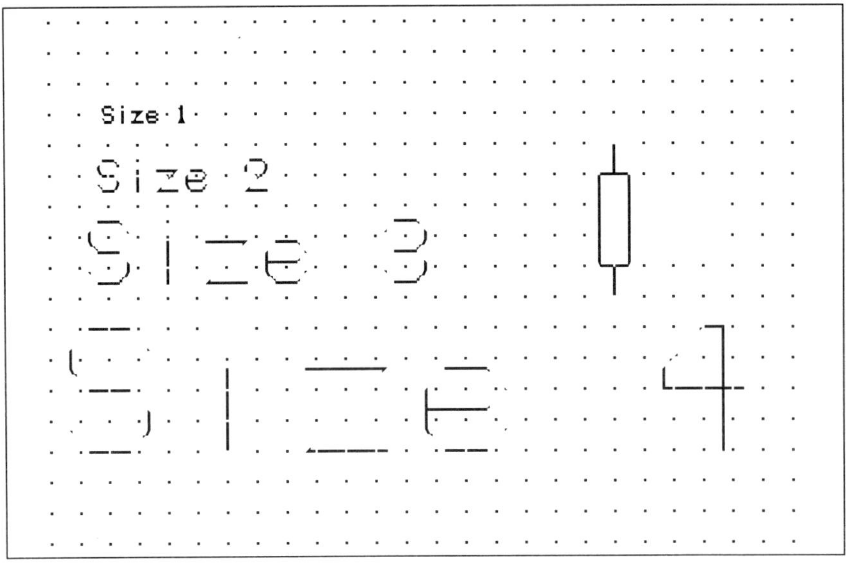

Figure 3.4 The size options for text – only the smallest size is useful for most component symbols, but the others are useful for tables and for naming a circuit.

The *Width* (key W) option allows you to change the thickness of the lines that are used for text. This is an option that you would use only for large print in headings, because the effect is coarse. The width numbers range from 1 to 8, and numbers above 2 are not suitable for small print. Figure 3.5 shows some widths as they appear on screen – note that the screen does not show the width differences well, and they appear correctly only on the printed copy. The widest lettering is also clearer on printed copy.

- Use wider lines only when you are also using larger sizes.
- Note that EASY-PC Professional allows you to change size and width by specifying figures rather than by selecting from a menu of fixed dimensions.

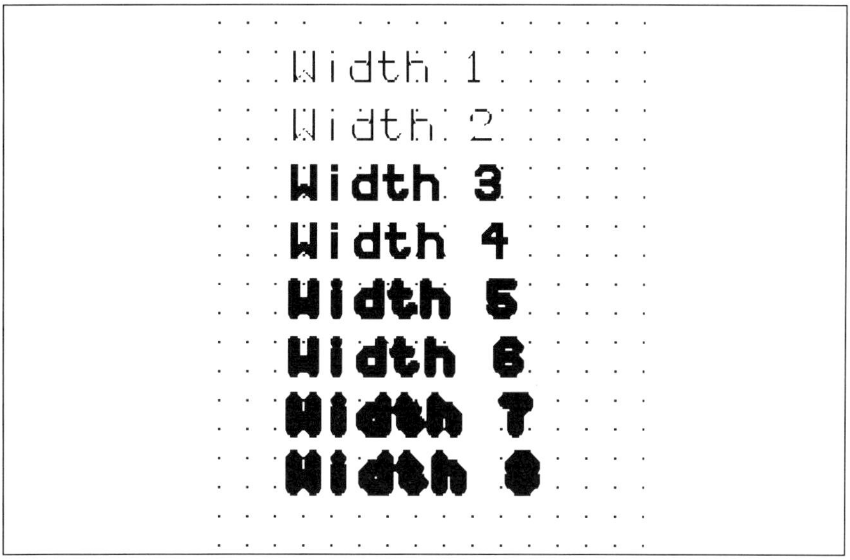

Figure 3.5 Text widths – width refers to the line used to draw text, not the overall width of a letter. The larger widths should be used only on the larger heights of text.

The *Angle* (A) option allows you to place your text along four possible baselines, using numbers 0 to 3. Angle 0 is the normal horizontal left-to-right lettering; Angle 1 is vertical, bottom to top, with the tops of the letters facing left; Angle 2 is upside-down lettering; and Angle 3 is vertical with the letters starting at the top and the tops of the letters facing right (Figure 3.6). This allows you to apply lettering along any axis of a schematic or a PC board.

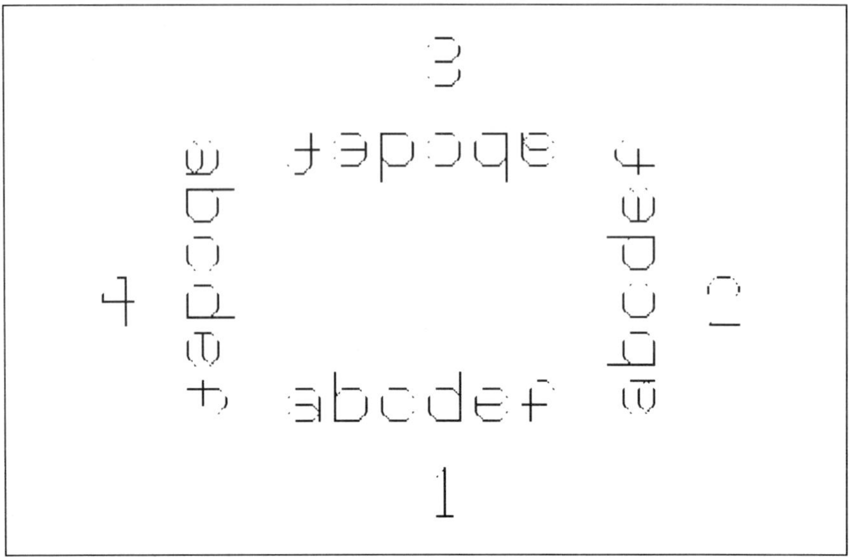

Figure 3.6 Using Angle to turn text around in 90° increments.

The *Flip* (F) command will turn a letter or phrase around on its vertical axis. This is not likely to be useful for schematic lettering but it can be useful on PCB lettering, where you want lettering reversed on the opposite side of a board.

The *Repeat* (R) command is used to make a copy of lettering which can then be put at some other part of a diagram by moving the mouse and clicking. This command is used in the same way as the Repeat command for a component symbol.

Finally, *Zap* (↑Z) is used to remove lettering. A Zap command is used in place of a Delete command where what is to be removed consists of several pieces, and you are always asked to confirm (using Y or N keys) that you really want to remove the item that the cursor is on.

Co-ordinates and cursor

EASY-PC normally appears with the cursor position co-ordinates in X,Y form using inches, arranged for a PCB of 17" x 17". You can switch this display to one of millimetres (or back to inches) by using the I key or by selecting Units from the right-hand square menu.

- You can fix this, and any other options that you currently have in force, by using User Defaults (↑F2) and selecting option A, to save the current defaults.
- You can also opt to use absolute or relative co-ordinates. Absolute co-ordinates are measured from the bottom left-hand corner (origin) of the PCB. Relative co-ordinates are measured from any point that you specify as an origin. The use of relative co-ordinates is particularly useful if you want to place a set of components at some fixed distance apart from each other. By making the relative origin at the origin of the first component, you can easily read off the positions for the others.

Another cursor option is size. The default cursor is a small one, and it is often easier to line up components if the cursor is large. Using the large cursor, as indicated in Figure 3.7, allows you to line up remote items, making the appearance of a schematic much neater. The long cursor line changes colour (from red to blue) when it is lying over an interconnection. This also can be made into a default by using the User Defaults A option.

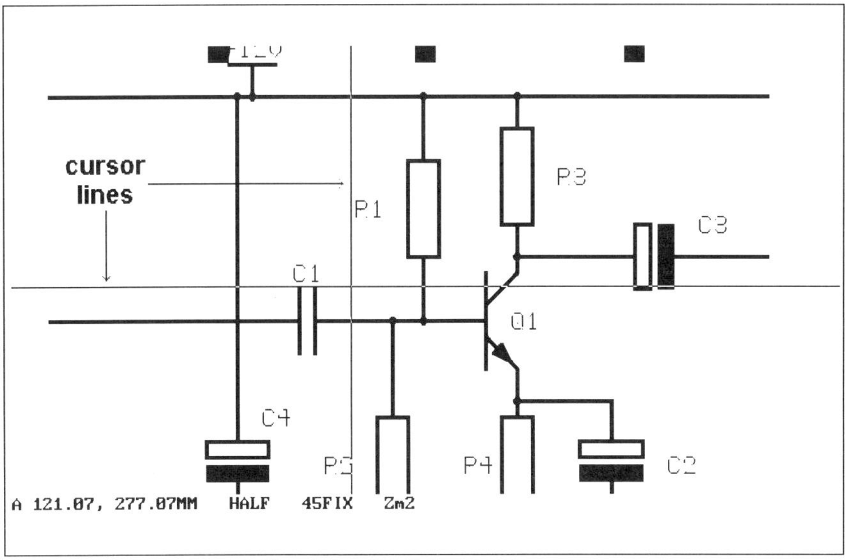

Figure 3.7 Using the large cursor. This is useful when you need to line up items that are some distance away. If you are lining up with something that is not currently visible, use the co-ordinates at the foot of the screen.

Block actions

A *block* has much the same meaning in EASY-PC as it has in word-processor and spreadsheet programs, an area that can be selected and worked with as a unit. When EASY-PC has marked some portion of a diagram as a Block, the marking is indicated by a rectangular box, Figure 3.8. This block has been positioned as follows:

1. Place the cursor where you want one corner of the box to be – the example has used the upper left-hand corner.
2. Press the F10 key – the cursor will change to a dot or small circle. Do not click the mouse.
3. Without pressing the mouse button, drag the shape out to a box that encloses the portion of schematic you want to use as a block.
4. Click to end the block marking.
   - If you want to remove the block marking use the ESC key or the right-hand mouse button.

Once a block has been defined in this way it can be used in actions that appear in the Block menu (Figure 3.9). First, however, you need to know

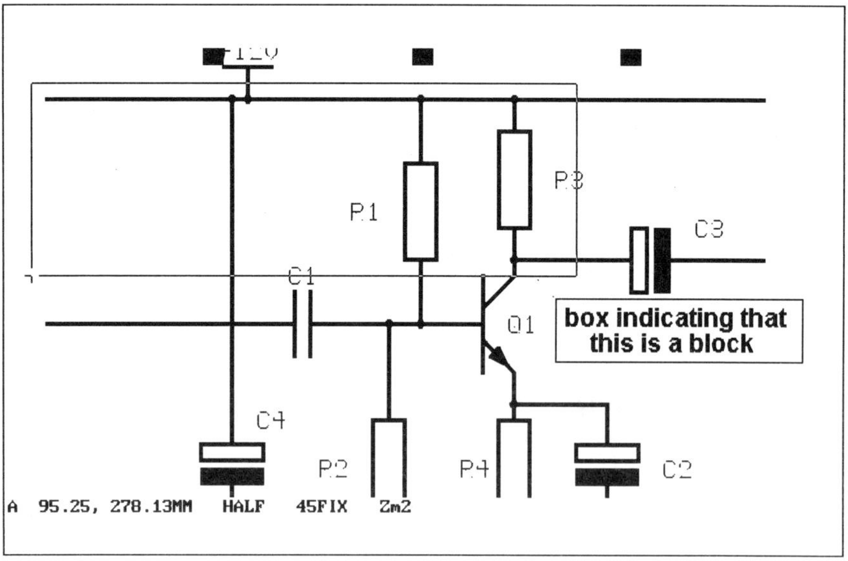

Figure 3.8 A portion of a diagram enclosed to mark it as a block. The Block editing commands now apply and are found in the central menu.

Figure 3.9 The Block menu as it appears on the screen. You would normally use the key options rather than draw down the menu.

what these block actions apply to. When you draw a block on a schematic, the block applies to components or interconnections whose origins are enclosed in the rectangular space. If the boundary omits the origin of a component, even if all the rest of the component is enclosed, then the block does not include that component, and the same is true of connecting lines.

- Always save your work before using a Block action. If the action does not do what you expected you can then recover the original arrangement.
- Another point to note is that the origin for the block marker itself is the point where you started marking the block. For example, if, like me, you always start marking a block at the top left-hand corner of the rectangle, then this will be the origin for your blocks. When you move the cursor and use the Shift command, for example, the origin of the rectangle will be placed at the cursor position. You can alter this default origin position by placing the cursor on some other point in the block and pressing the O key.

Shift (S) will move the whole block to the cursor position. What actually moves to the cursor position is the origin or base-point, which is the starting corner of the rectangular block marker, unless you have shifted the origin using the O key (see above). You can use this Shift command to move a complete piece of drawing into another position, perhaps to make room for another part of a diagram.

Repeat (R) will make a copy of a marked block, placing the copy at the cursor position. As before, it will be the origin or base-point of the block that will be placed at the cursor position. This can be useful for diagrams in which there is a repetitive structure, such as collector loads in a transistor schematic or complete units of digital IC schematics.

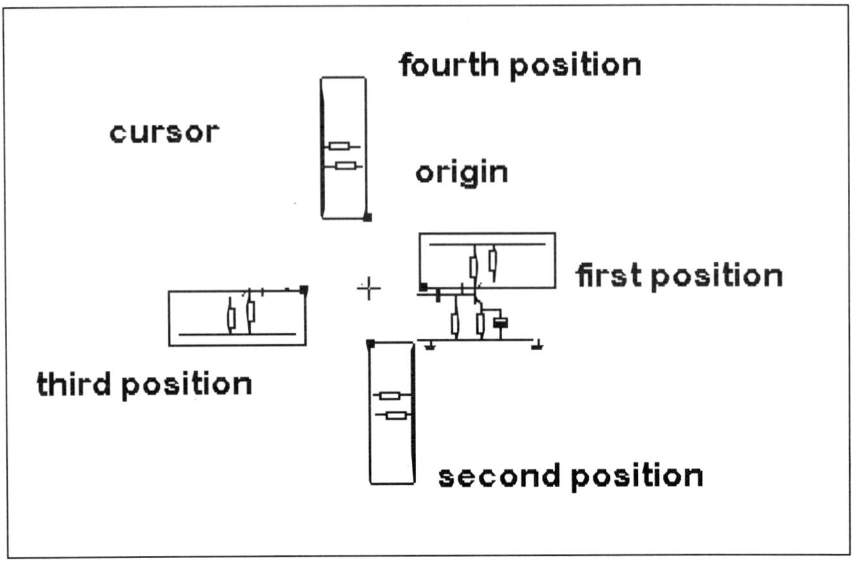

Figure 3.10 Using the Rotate option on a part-drawn circuit. The rotation is around the cursor position, and the origin of the block is kept at a constant distance from the cursor mark.

Clockwise (C) and *Anticlockwise* (A) are used to rotate a block in the directions indicated. The rotation uses the cursor position as a centre, with the origin of the block rotated, Figure 3.10. The origin has been deliberately marked with a small cross in this composite diagram of a part-drawn circuit to show its rotation around the cursor position.

- If the cursor is located at 8.5, 8.5 (using inch units and absolute co-ordinates) it is always possible to rotate a block. If you cannot rotate a block, put the cursor at this position first.

The *Flip* (F) block command will mirror a block with respect to a vertical line drawn through the cursor, Figure 3.11. This is useful for mirroring transistors or gates for some types of drawings such as multivibrators, which need symmetrical placing of components.

The *Draft*, *Normal* and *Bold* commands are used in connection with printing, using a dot-matrix printer, and have no application as far as the normal use of the screen images is concerned (see later).

The *Zap* (↑Z) command is used to erase a block. You need to use this with care, and you will be asked to confirm its use with the Y or N key. Remember that if you zap a drawing or portion of drawing which you have not saved it will be irrevocably erased.

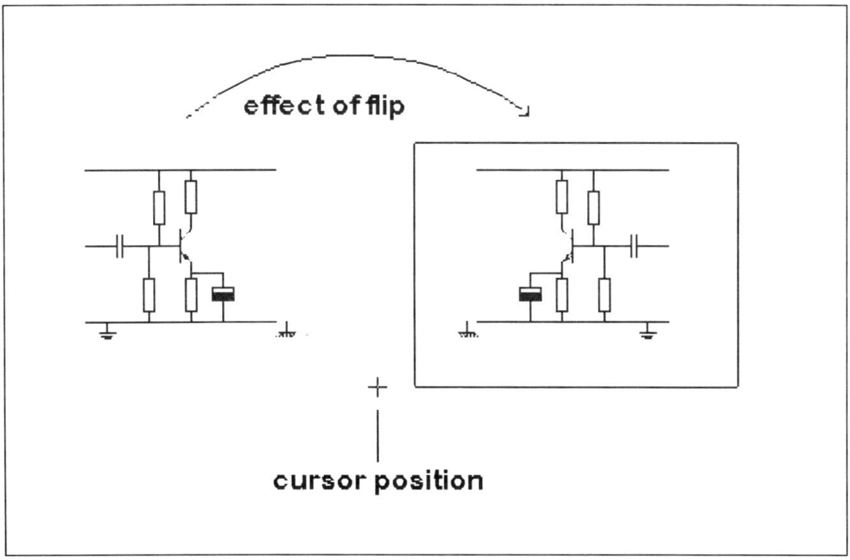

Figure 3.11 Using a Block Flip to make a mirror image of the items contained in a block enclosing a part-drawn circuit.

Printing your schematic

Printing your schematic is the most important action of all, because this is the whole objective of using EASY-PC to draw the schematic in the first place. The main EASY-PC program contains routines that allow for printing on a dot-matrix printer, but many PC users nowadays will have opted for a laser printer on grounds of better quality, easy use of single sheets of paper and reasonably low running costs.

- The actions for printing with a laser printer apply also to any H-P ink-jet printer or any ink-jet printer that is fully compatible with the H-P type. If you experience problems with an ink-jet you will need to check that your printer driver allows a long time for transmission retry – this is easy to set with Windows Control Panel if you are working from Windows. If you are working from DOS you may be able to set switches in the printer for an indefinite transmission retry. This avoids the printer shutting down while it is waiting for data.

Printing an EASY-PC file on a laser printer, or H-P compatible ink-jet, makes use of a supplementary program called EASYLASE. To use this you have to leave the main EASY-PC program after saving your schematic as a file. You then start up EASYLASE, open the file and print using the commands of EASYLASE (see later).

- If you are using EASY-PC from Windows you can switch to EASYLASE without shutting down EASY-PC – but you need to ensure that the file you want to print has been saved.

Printing on a dot-matrix printer is done as follows:

1 Ensure that the printer is connected and on-line, and check its type as 9-pin or 24-pin, narrow or wide carriage, and using 1⁄216" or 1⁄180" line feed intervals.

2 You create a block of the schematic as seen on screen. This can include a table of components, or as much blank space as you want around your diagram.

3 You can select one of the printing preferences: Normal (N), Bold (B) or Draft (D). You should *always* use Bold for the final copy.

4 You will see the User Defaults menu, Figure 3.12, with the printing options arrowed.

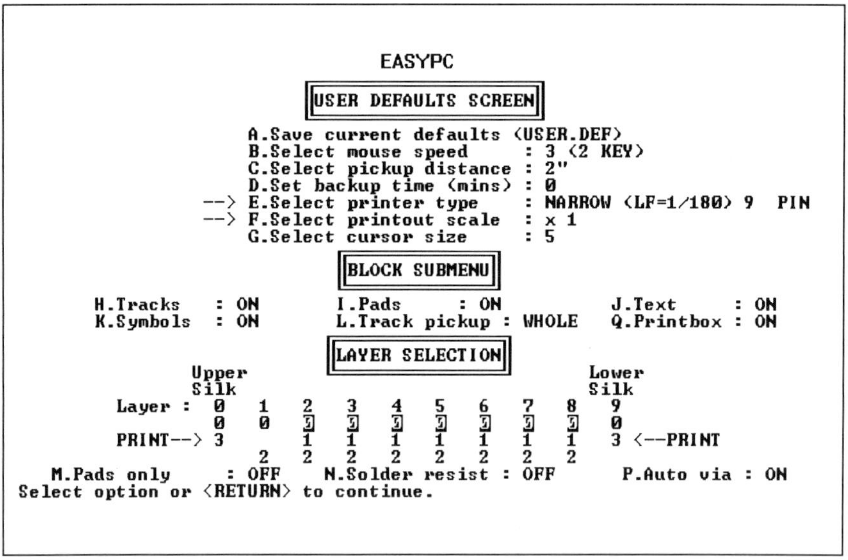

Figure 3.12 The User Defaults screen, illustrating the Printer defaults for a dot-matrix printer. Note that this does not apply if you use any other type of printer.

5 From this menu use the E key to select the type of dot-matrix printer you are using. The types are 9-pin or 24-pin, with a line feed distance of 1⁄216" or 1⁄180", and with a wide or a narrow carriage. There are six choices in all (there is no 1⁄216" option for 24-pin printers), and pressing the E key will cycle you through the options.

- Note that since you will probably not change your printer often, you can set this option and then use the A key to save the defaults in this menu.

6 The print scale must also be chosen. The default is 1 and using the F key will cycle you through the options of ½, ¼, 4, 2 and 1 again.

- If you try to print at too large a scale (which would make the print area larger than the paper) you will be returned to the diagram with a warning that you are trying to use too large a scale.

When you select a usable scale you will see the screen version of the schematic return and the word `PRINTING` appears over the diagram. This does not guarantee that printing will take place, because this message will appear even if there is no printer connected. You will need to use the CTRL-ALT-DEL key combination to escape from this, losing any file that you have not saved.

If you run EASY-PC from Windows and encounter this problem, you can use the CTRL-ESC keys to escape to the Program Manager. You can then click on the EASY-PC icon which will be visible at the foot of the screen (you may need to move other windows out of the way). The small menu that appears contains a Settings option, and when you click on this you will see a window of other options, of which one is *Terminate*. Click on this option and on the OK button that follows it, to escape from EASY-PC. You can then re-start EASY-PC without the need to leave Windows or re-boot the computer.

- Another option, if you have Windows correctly set up, is to use CTRL-ALT-DEL. This will allow you to escape back to Windows (a second CTRL-ALT-DEL will re-boot the computer).

Using EASYLASE

Your laser printer must be fitted with adequate memory. If only 0.5 Mbyte is fitted, only part of a page will be printed at 300 dots per inch, though you might be able to print a full page working at 150 dots per inch resolution.

Users of laser printers or H-P compatible ink-jets need to print from a saved SCH file – note the filename of the schematic before you change programs. If you are running from DOS you must then shut down EASY-PC by returning to the main menu (↑F10 key) and selecting item 5, the *EXIT to DOS* option.

You then need to start the EASYLASE program. If you are running these programs from Windows (so that you do not need to shut down EASY-PC), EASYLASE must be run, like EASY-PC, full-screen, *not* inside a small window. Unless you installed it to a different directory its full name and path will be:

```
C:\EASYPC\EASYLASE.EXE
```

When EASYLASE starts, you will be asked for the filename of the drawing that you want to print, and the default of NEWBOARD.PCB will be offered. You will need to type the full name for your file, such as AMP1.SCH – you must use the extension.

- You can combine the program command with the filename using a line such as:

```
C:\EASYPC\EASYLASE AMP1.SCH
```

When the file is loaded in to EASYLASE, you will see the main control panel, Figure 3.13. This contains sixteen options that you can select by using letter keys.

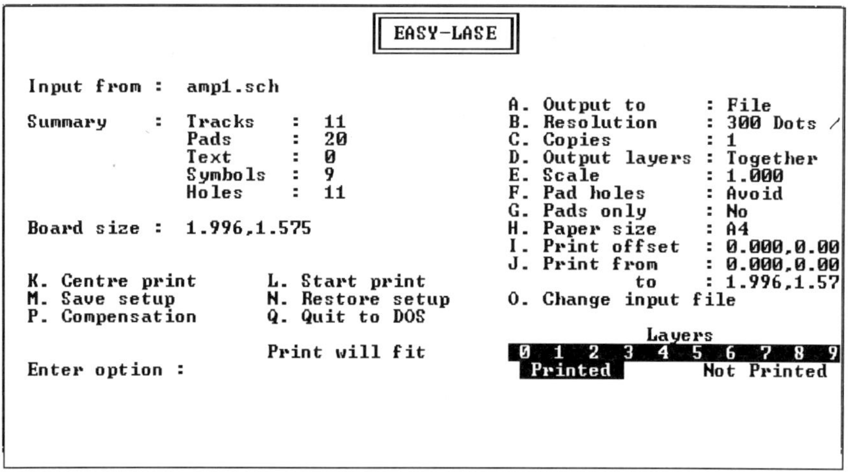

Figure 3.13 Using a laser printer. The EASYLASE program will read an EASY-PC file (schematic or PCB) and print it, and this is the default menu screen, which offers printing to a file.

A allows you to switch between output to File, to COM1 (main serial port), to COM2 (second serial port), to LPT1 (main printer port) and to LPT2 (second printer port). The File output is the default the first time you use EASYLASE, after which it will use the port that you specify. You should save the new defaults after you have made any changes to the settings.

- You must not select a port that is not physically present – use LPT1 unless you are certain that your printer is connected to some other port. Use *File* if you want to print the file on a printer that is connected to another PC.

B allows you to select resolution, with the default at 300 dots per inch. This produces diagrams of good quality but for many applications you will get perfectly satisfactory output at a much faster rate by using a lower resolution. Pressing the B key will switch among 300, 75, 100 and 150 dots per inch.

C determines the number of copies. When you press this key, you will see a message at the bottom left-hand corner of the screen, and you can delete the default of 1 copy and type in the number that you want to print.

- For a new drawing, always print a single copy first to make certain that it is as you want it.

D allows you to select output layers, with a default of *Together*. For schematics this is the only option – you have other options for PCB layouts (see later).

E is the important scale selection, set initially at 1. When you press the E key, you can type in the scale you want to use, and the screen message will be either `Print will fit` or `Print will NOT fit`. You need to use some trial and error on this to find the appropriate setting – in the example, the largest whole-number usable scale on A4 paper was 3. You can, however, use scales such as 3.2, 3.44 and so on to get a fit to almost the full size of the paper.

- If you make the scale too small, or forget to select the schematic as a block, you may find the resulting print too small to read.

- On most diagrams which use more width than depth, it is an advantage to rotate the drawing (see later).

F is the *Pad Holes* option, used for PCB work. The options are *Avoid* or *Fill*.

G is the *Pads Only* option for PCB use, with only the *No* option available for a schematic.

H is the important paper size option, initially *A4* and which can be switched to *EXECUTIVE*, *LETTER* or *LEGAL* sizes.

I allows for print offset in X and/or Y directions. When you use this key you will be asked to enter the offset dimensions in inches: see below.

J confirms the block selection by showing the dimensions that are to be printed. In the example, these are shown as from 0.000,0.000 (there is no offset figure specified) to 1.996,1.575, indicating that a 1:1 scale print would be 1.996" wide and 1.575" deep. This is why a 4:1 scale was not allowed in this example – there would have been insufficient margin on A4 paper.

- Checking this size figure can be a useful way of estimating the scale that you can use, eliminating guesswork.

O allows you to change to another file for printing, either before you print this one or after.

K is a useful option, to centre the printing on the paper. Pressing this key will automatically enter an offset, and if you want to return to printing starting at the edges you need to use the *I* key and reduce both offsets to zero again.

L is used to start the print action after you have checked that all of your options have been correctly selected.

M allows you to save the selections that you have made so that they will become defaults for the next time you use EASYLASE.

N will restore the defaults that have been most recently saved, if you are not satisfied with changes that you have made and not saved.

P will apply compensation to allow for imperfections of your printer. This is used mainly for PCB work if the printouts do not line up perfectly. You may find, for example, that irregularities in the printer roller make distances on the printed diagram vary slightly. These variations will be systematic, meaning that they recur at predictable intervals. See the end of this chapter for details of using the compensation process.

Q allows you to Quit back to MS-DOS (or to Windows if you launched EASYLASE from Windows).

When you use a printer (as distinct from a file option), you will see a count (0 to 100) taking place of Tracks, Pads, Text and Symbols. This can take some considerable time, and the printing of a schematic will not start until the Tracks, Text and Symbols count is complete (the Pads count is very fast on a schematic). The time can be quite long even for a simple file, and if you are printing something like the TVIF sample file you can take a coffee break (possibly a lunch break unless you have a fast PC). The symbols count takes more time than the others.

- When you use a file option, you will be asked for a filename, which can include a floppy disk drive letter or a hard disk directory path.
- You should always save your settings once you have established what you need, because this can save a lot of time in subsequent schematic printing work.
- If you are likely to need more than one copy of a schematic, you should opt for printing several copies, because the time to print ten copies, for example, is very little more than the time to print one copy. In

addition, if you run EASY-PC from Windows you can set up the program (see any good book on Windows) to allow the printing to be in the background, allowing you to get on with other Windows tasks.

See Appendix G for the use of plotters in printing either schematics or PCB layouts.

Creating and saving symbols

The symbols that you are provided with are fairly comprehensive but you may need additional symbols, such as the zig-zag resistor symbol, and EASY-PC provides for creating these for yourself. Creating a symbol for schematics uses option 4 (*Create Schematic Symbol*) on the main menu.

This takes you to a screen that looks much the same as that used for a schematic, and you might wonder where to start. A good start is to zoom to the size that you normally use for schematics work, and it helps enormously to load in an existing symbol for size comparison. You will, for example, need to ensure that you use the same track width for the connecting lines to the new symbol.

Now select the New Track key, F2. When you have done this, clicking on the central menu square will reveal a new menu, Figure 3.14. This is essentially the same menu as is used for drawing tracks. Some functions, like *Repeat* and *Zap*, will be familiar from previous use in other menus. The first four commands are used in relation to lines drawn on the screen.

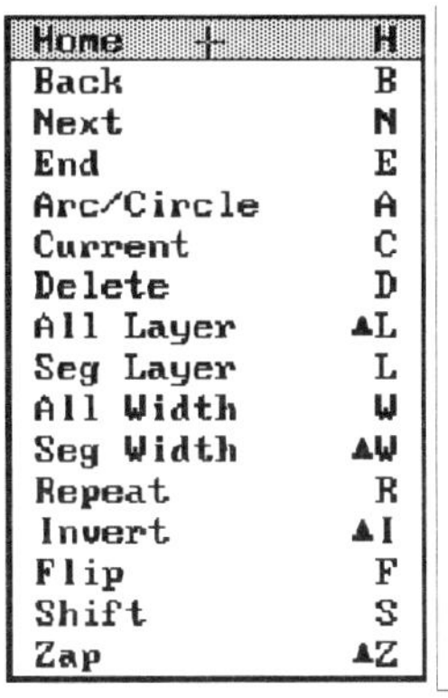

Figure 3.14 The central menu of options for track editing.

Because drawing a symbol uses the instructions for drawing a track, the same rules apply, but as the commands are not obvious, first try drawing a simple + sign on the screen. The method (Figure 3.15) is as follows:

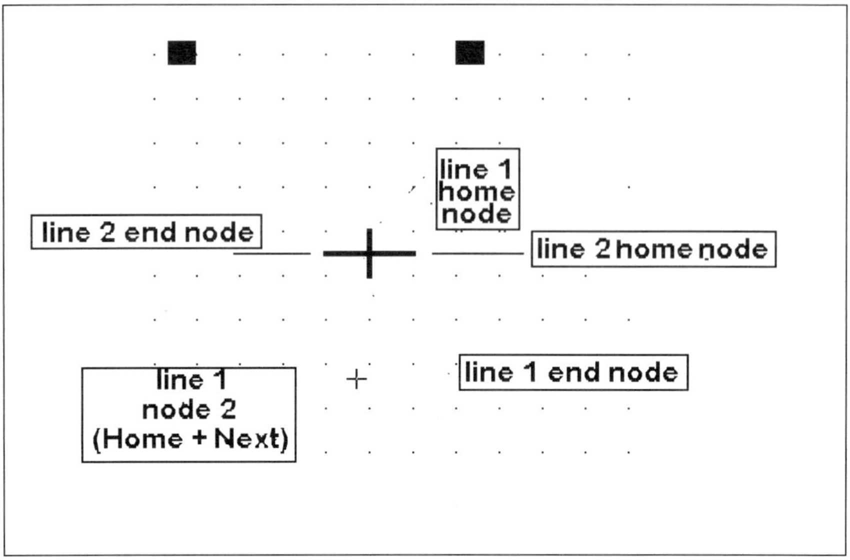

Figure 3.15 Creating a cross symbol to add to the Symbol Library. This is a useful exercise in drawing which provides skills that you will later find useful.

1 Press the ESC key until red diamonds appear on the status line, to make sure that no commands are engaged and that you are not in an edit mode. Set the snap to half of the spacing between grid dots, and switch on the grid. Work at a large magnification – the illustration shows zoom level 2.

2 Place the cursor on the point that will be the start of the vertical line of the cross. This is the home node for this first line.

3 Press F2 for New Track. Move the cursor to where the end of this vertical line will appear and click. The line will be visible.

4 Now use the ESC key or the right-hand mouse button to escape. Place the cursor at the home position for the horizontal line. Press F2 and RETURN (or the left-hand mouse button) and draw the line as before. Escape again, and the + sign is drawn.

   - Remember that you must use the ESC key or right-hand mouse button to end the line-drawing action.

The rules for these lines are strict and the same as those for schematic lines or PCB tracks.

- A line starts at home node and ends at end node. Unless you escape from the line, any further extension to the line is made from the end node, nowhere else.

- You can click the mouse button (or use the RETURN key) at several points on a line, while you are drawing it, to make nodes. You can change the shape of the line by using these nodes.
- If, during line editing (F1 key), you click to one side of a line near a node, the line will change so that the new point becomes a node on the line. The line now joins this node with the others, excluding the original node which you can then delete if you want to.
- You should practice the effects of using the B and N keys while in line edit mode.

Figure 3.16 show several steps in the making of an arrow shape, indicating where the mouse button has been clicked and when the node has been shifted.

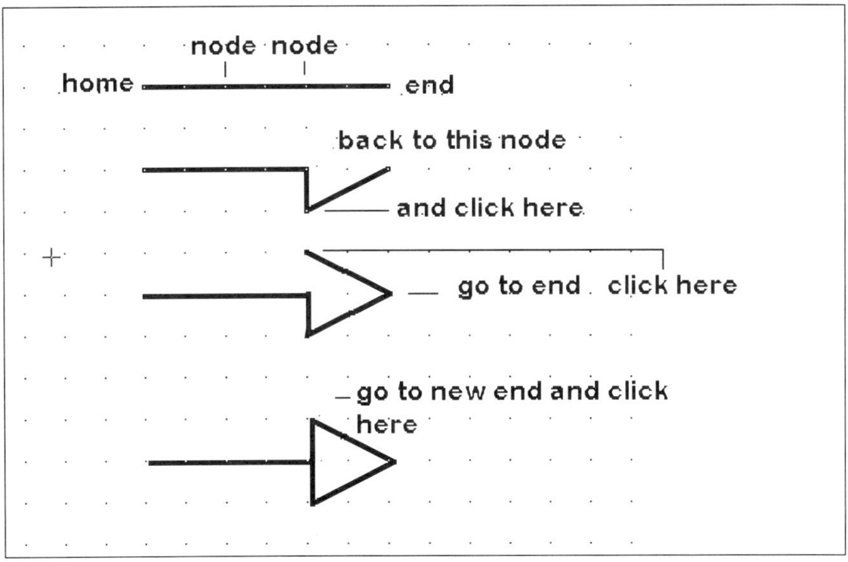

Figure 3.16 The steps in creating an arrow shape.

- If you find that an extension to a line is not as you expect it, usually with a crook in the line, go to the unwanted node and use the D (Delete) key. Figure 3.17 shows a common problem, that the second angled line is drawn so that the angles are 90° and 45° only. You can tackle this by going back to the unwanted node and using D, or setting *Multi-angle Fix* before you start drawing.

Practice making the arrow shape. Once you can confidently produce shapes like this you will have few problems using EASY-PC for symbol construction or in using it for PCB tracks.

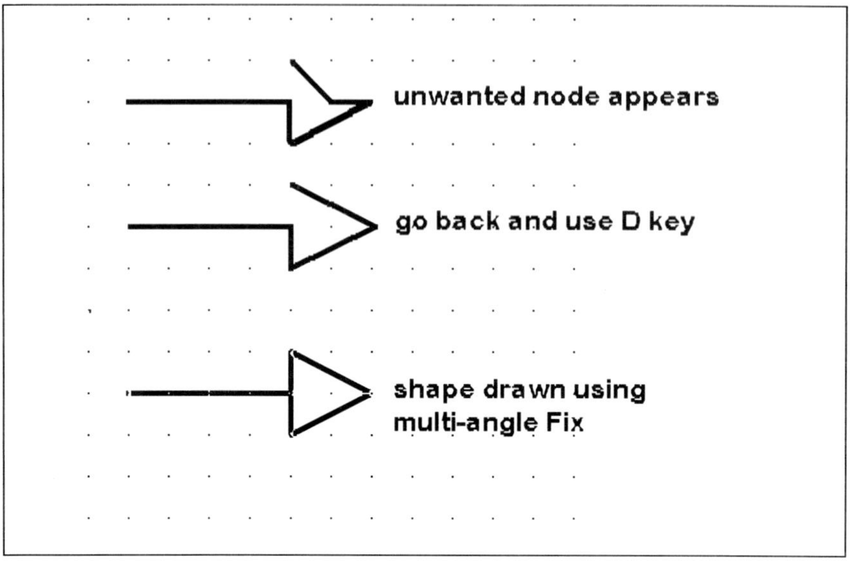

Figure 3.17 How to edit out an unwanted node so as to correct a mistake. Use rubber-banding to avoid the worst errors.

Shapes that you might want in creating symbols are likely to need circles and arcs. Once again, you need to be careful about your use of nodes, because a circle always starts with a two-node straight line, which is a radius of the circle and which has only the home and end nodes – you must not click anywhere else to put in an intermediate node.

- Remember that you can put a block around a bad drawing and use ↑Z to zap it.

Figure 3.18 shows a line and the circle that can be obtained from it (after shifting it down and using the A (Arc/Circle) command). When you press the A key with the cursor on the end position of the line, you will be asked to choose tolerance in terms of Low, Medium, High or Auto. If you simply press the RETURN key or the left-hand mouse button, you will use the *Auto* setting, which is the most suitable for most cases.

- The Auto option is correct for most work. If you opt for High, your circles will be very precise but they will use a large amount of memory and will take some time to draw. If you use high resolution on work that contains a large number of circles, like a panel layout, you may run out of memory. By contrast, using low resolution may provide you with circles that look too angular (polygonal) for your liking. The Auto setting uses the lowest resolution that will provide an acceptable appearance.

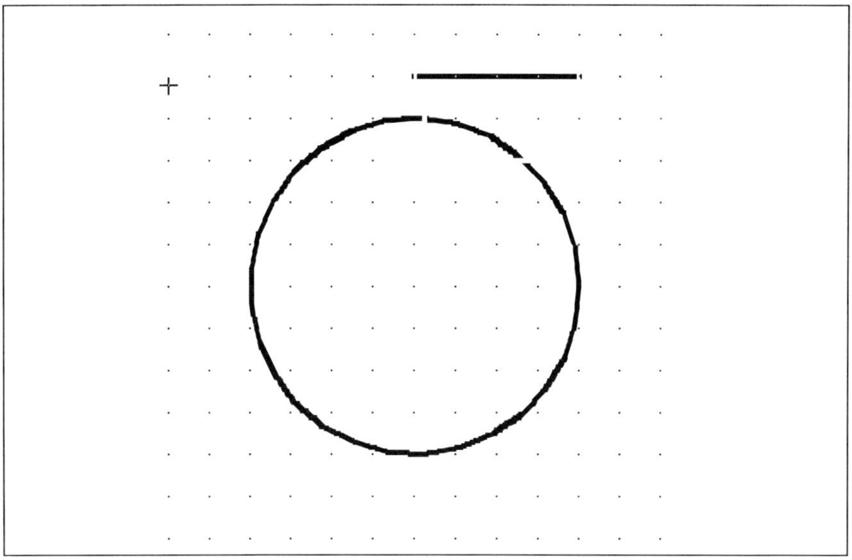

Figure 3.18 Illustrating a line, which is the first step in drawing a circle. The line is a radius of the circle, and the **A** key will convert this line into a circle that is made up of short segments.

A circle is changed into an arc by deleting segments. When EASY-PC draws a circle it does so by drawing a set of short straight lines with close-spaced nodes. How it does so depends on where you start from.

Figure 3.19 shows a circle which has been drawn and has had segments deleted. The home position has been selected, and the D key pressed several

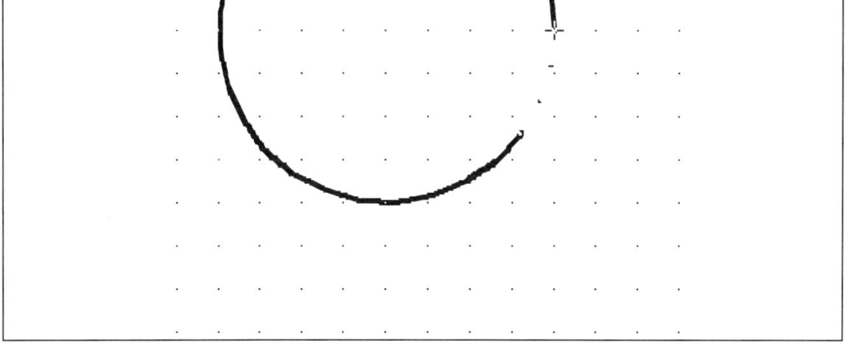

Figure 3.19 Segments deleted from a circle to form an arc.

times. Each time the D key is used a segment is deleted, leaving an arc. When you start at the home position, as illustrated, the segments are deleted in a clockwise direction starting from the home position. If you start by selecting the end node, the deletion is anticlockwise starting at the end position.

- Both home and end positions appear identical to your eye. You can select them only by using the H and E keys.

Saving a symbol

Figure 3.20 shows the addition of a circular outline to a transistor symbol. The symbol called N has been inserted, using Zoom 7, and a track line has been drawn part-way down the line of the base. The A key has been used to draw the circle, and the cursor has been moved and the S key pressed to shift the circle until it is correctly placed.

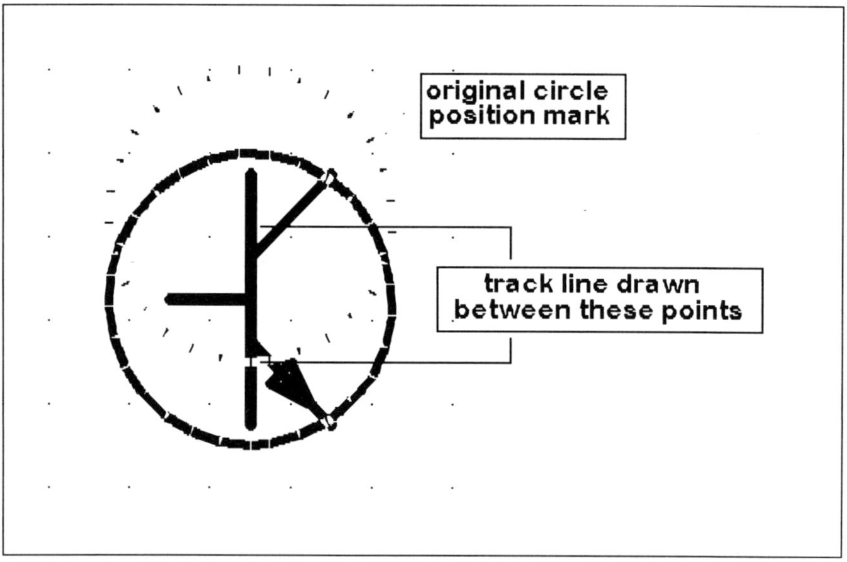

Figure 3.20 Making a circular frame for a transistor symbol.

- Note that by using a standard symbol, the track width has been set to match the track width on the existing symbol.
- Note that this leaves a trace of the original nodes of the circle as it was first drawn. Use the Pan command to refresh the screen.

This new symbol can now be saved. First you must escape from the last drawing command you have used, and define the origin or base-point of the new symbol. In this case, place the cursor at the end of the base lead, which is the point used for the skeleton transistor symbol, and press the O key or use the Origin command from the menu.

When you have defined the origin (check by moving the cursor and pressing ↑H, which should put the cursor back to the origin), use the left-hand square menu (or press ↑F5) to get the Symbol Library. From this menu (available on the central square) you can select from:

Dir *.SLB, key D, to show the library contents
Add to Library, key A
Scan Library, key S
Erase item, key E
Update definition, key ↑D

In this case you need to press the A key to add this symbol to the library. You will be asked, as a reminder, if you have set the origin, and if you have and reply with the Y key, you are then asked to confirm the symbol library name. The name SYMB.SLB will appear as a default.

- Preferably you should not use this default name. The reason is that this is a standard library which at some time in the future you may update by obtaining a new library of this name from Number One Systems. If you do so you will lose this new symbol because it will not be in the new library. You should preferably use a name of your own for a library of symbols that you will create, such as MYSYMB.SLB.

You are then asked to type a name for the symbol, and since N is used for the skeleton *NPN* symbol, it seems appropriate to use NPN for this one.

From this point on you can obtain from this library a transistor symbol with a circle around it. You can create other symbols, either by using existing symbols or by starting from scratch.

- For a dot, use a pad with its hole filled in.

You might, if you need diagrams for the USA or for some journals, need the zig-zag resistor symbol, and this is also relatively easy to make. You might also like to load in symbols and rotate them, saving the result as another symbol (such as *RH* and *RV* for horizontal and vertical resistor, respectively).

The remaining commands of the track-drawing menu contain several that are particularly intended for PCB layout work rather than for symbols for schematics. You can, for example, ignore the *All Layer* and *Seg. Layer* commands.

- The *All Width* and *Seg. Width* commands allow you to select a width figure for all of a track line or for the single segment that follows the current node. This can be used if you find that your symbol is using lines that do not match the connecting lines for other symbols.

The *Repeat* command is used to copy a line or a selected block to the cursor position, as we have noted earlier. *Invert* will swap the positions of home and end, to allow you to draw an extended line from what was the home position (a line can be extended only from the end node).

The *Flip* (F) command will reverse a symbol, and can be used in schematics where transistors are shown in mirror-image positions (such as a flip-flop). You should not use Flip on PCB components, particularly ICs, since this would invert the pin positions.

- If you need to invert an IC position, rotate the symbol by 180°.

Shift (S) will move a symbol in the usual way and *Zap* (↑Z) will delete the symbol shape completely. Remember that you can also use a block outline and the angular rotation numbers along with Flip to change one symbol into another that is differently oriented.

Printer compensation

1. Start EASYLASE and load in the file LASECOMP.SCH, which is supplied as part of the EASY-PC package.
2. Print the file using 300 dots per inch resolution, scale 1.000 and the compensation factor (P key) set also to 1.000.
3. The image consists of a grid whose top line and right-hand line are sloping so as to create a Vernier effect, Figure 3.21. The horizontal centre line should measure exactly 5.000 inches and the central vertical line should measure exactly 10.000 inches.
   - These dimensions will be changed if you have to print at a reduced scale. For example, if you have to print at scale 0.8 then the dimensions will be 0.8 times these amounts: 4.000 inches and 8.000 inches.
4. Using a finely-graduated steel rule, measure the horizontal line on your printout. If the result is not 5.000 inches, move the position of the rule (up or down) until you find a position where the reading of width is exactly 5.000 inches (or whatever dimension is provided by the scale you are using). Where the rule cuts the left-hand scale on the grid will show the X-axis compensation factor – each small division corresponds to a compensation factor of 0.0002.

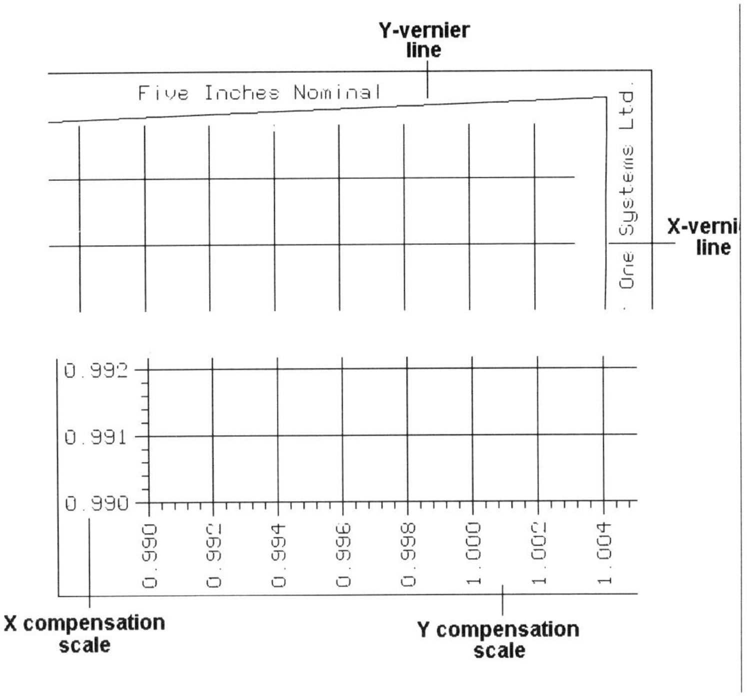

Figure 3.21 Views of opposite corners of the LASECOMP.SCH pattern, which contains vernier scales for measuring compensation factors for a laser or H-P compatible ink-jet printer.

5 Now measure the central vertical line, and if this is not exactly 10.000 inches, move the rule sideways until you find where the measurement is exactly 10.000 inches (unless you have printed with a scale factor other than 1.000). Read the Y-axis compensation from the scale on the bottom of the display. On this scale, one small division corresponds to a factor of 0.0004.

6 Return to the EASYLASE menu and use the P option to enter the X factor followed by the Y factor. Save these as defaults by using the M option.

7 Don't assume that these compensations will be once and for all figures. For the most precise work you should measure and apply the compensation on the day that you are printing the work that must be particularly precise.

NOTE: Additional libraries of symbols can be obtained from Number One Systems on request.

4 PCB drawing

Outline

Preparing a PCB layout using EASY-PC follows the same general methods as are used for a schematic, but with the complications of layers (Figure 4.1). EASY-PC allows for up to two silk-screen layers that will be used for printing (usually in white) on the surfaces of the board. In addition, tracks can be placed on the underside and also on the silk-screen side, a total of four layers. For multi-layer boards, a total of up to ten layers (two silk-screen and eight copper track) can be used if you can think in all these dimensions.

The layer defaults are important. When you start EASY-PC running, it will use layer 1, which is the main track layer whose tracks are shown, on the default colour scheme, in red. The silk-screen printing is shown in white, and layer 8 is shown in blue. All mounting pads that make contact through all layers are shown in white. If you want to alter the defaults it can be done,

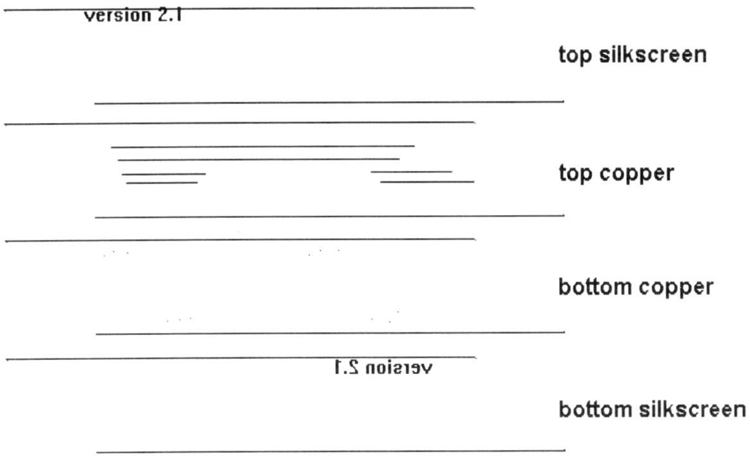

Figure 4.1 The simplest set of layers – a board with both sides coppered and silk-screened. This can be thought of as four layers, two of silk-screen lettering and two of etched copper for connections.

but you should retain this set of defaults until you are experienced in the use of EASY-PC.

- For surface-mounted components, pads will normally be placed on one surface only and the normal default is to have such pads shown in red or blue, depending on which layer they are on.

When you design a layout using EASY-PC you always look at the board from the top (silk-screen 0) side. This applies to all of the layers, so that when you look at the pads and interconnections that you have made on the rear copper-track side of a double-sided board, printing out this diagram will show these pads and tracks the wrong way round. This has been designed for convenience, because it is easier to design your tracks if you are working from the same point of view for both front and rear. This arrangement is also suitable for sending output in the form of Gerber and Autodrill files when you send files to a bureau for conversion into film masters and drill patterns respectively. If you are using the normal DIY expose and etch methods then when you want to print the layout, you can use Block Flip on the PCB, so the layer that represents the tracks on the back side will be printed in the way that can be used for etching the board.

- The examples that are included in the manual show all layers as seen from the front, or silk-screen 0 side, of a board, as would be used for files sent to a bureau. A list of PCB preparation bureaux is provided in the manual for EASY-PC.

The symbol library for PCB shapes is named SYMB.LLB, as distinct from the much larger SYMB.SLB library that is used for schematic symbols. SYMB.LLB is selected automatically when you take the PCB options from the main menu. The library contains patterns for surface-mounting components as well as the older type, and includes connector sockets and pin-strips. A complete Eurocard shape is included as a single component, complete with connectors. Figure 4.2 shows an example of the Eurocard silk-screen loaded with two 16-pin IC positions – this is FIGURE4.PCB of the EASY-PC sample files.

PCB design also calls for the use of pads and vias. A *pad* is a connector to a component, and all-layer pads are the most common type, allowing connections to be made to a component from either side of the board. Single-layer pads are used for SMT layouts. The *via* is a connector that goes through the board to connect a track on one layer with a track on another layer. For the sake of simplicity, we shall assume that the most complicated board you want to use has copper layers front and back, with the silk-screen printing on the front, so that a via would join a track on the front with one on the back.

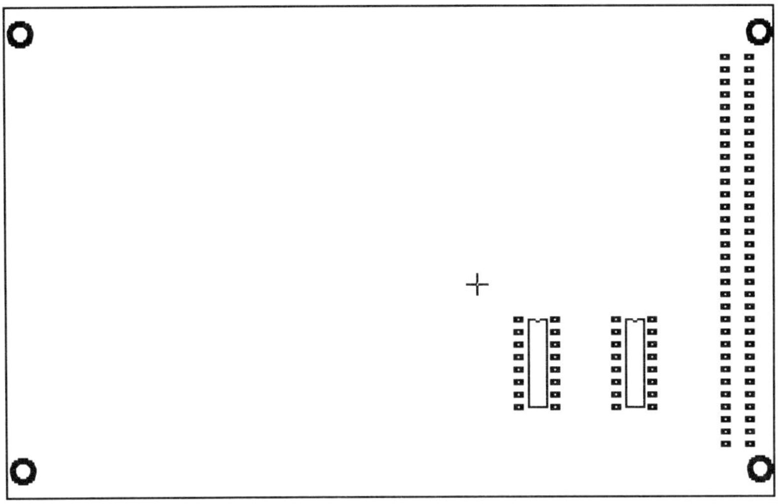

Figure 4.2 The standard Eurocard PCB layout with a couple of IC holders in place.

In addition, track width can be varied, so that you can create wide tracks for earth planes, and for busbars that will carry a larger current than is normal on the narrow type of track. Conversely you can select very narrow tracks to sneak their way between pins of an IC without touching.

- The most common track width for crowded layouts is 0.010". If you are etching your own layouts, you should aim for track widths of 0.050" or more, because the crude etching processes used for DIY work are not capable of achieving fine lines with any degree of reliability. If you are using a PCB bureau you must remember that the cost of producing a board will be much greater if you insist on using narrow tracks, and in this respect a width of 0.006" should be taken as an absolute minimum.
- For single-sided boards, a minimum width of 0.015 " is better because these boards are usually of a cheaper base material, and they are normally manufactured using the lowest-cost processes. This dimension allows 0.010" clearance between pins – remember that these dimensions will change on etching, giving about 0.013" and 0.012" respectively.
- If a layout would require you to try to take two tracks between a pair of IC pins you should put in a hard-wired link. The cost of adding the wire link will be less than the wastage involved in trying to make boards with such narrow tracks.

A simple example

An example is far more useful than notes about the way that EASY-PC creates a layout. It is also much easier to see what is going on when the example is a simple one, so we'll use the simple transistor amplifier circuit that has featured in the schematics of Chapters 2 and 3. Work on a nine-component circuit like this is easier to follow than on a circuit such as the examples that are provided with EASY-PC.

To start with, you need to have a schematic printed with components labelled, as in Figure 4.3. These references can be used on the silk-screen pattern, and you will label the transistor connections also on the silk-screen, remembering that you are looking from the top of the components, not (as is usual for component pin numbering) from the underside. You can usually make a reasonable guess on the size of board you will need at this point, but this does not commit you to anything because provided that your board requirements do not exceed the 17" x 17" limits, you can show cutting lines in the form of a border of silk-screen lines around your board area.

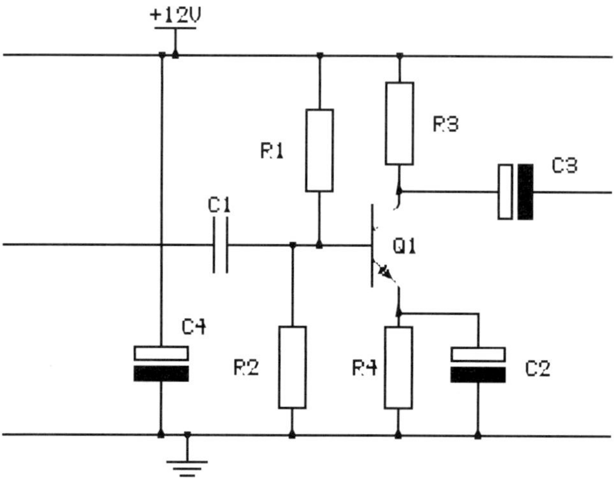

Figure 4.3 The schematic as it will be printed, with components labelled.

- This border should be drawn so that it is at least 2 mm beyond the outermost tracks.
- This will ensure that when the diagram is printed nothing will be missed. If a schematic contains isolated lettering (such as a version number) then there is a chance that this might not be printed because

it has not been detected as part of the schematic. Drawing a border eliminates this possibility.

The next stage is to place the symbols on to a part of the board. This will be done using layers 0 (silk-screen and component side) and 8 (track side) on a single-sided board.

1 Select User Defaults. The lowest part of the display shows a list of layer numbers from 0 to 9, and under them a set of selection indicators and numbers that are codes for colours (1=red, 2=blue, 3=white). By typing a number key you can determine which layers are active and what colours will be used for tracks (white is normally used for silk-screen and to show tracks crossing, but you can alter this colour scheme to suit your own purposes).

2 Tap the number keys until you reach the settings of layer 0 on, layers 1 to 7 and 9 off, and layer 8 flashing colour code 2 (Figure 4.4).

- This is not a normal arrangement, but there is no harm in using layer 8 for the rear side of a board when layer 1 is not being used for the front side, provided you remember that you are doing so – the point is to show that you can select layers for yourself.

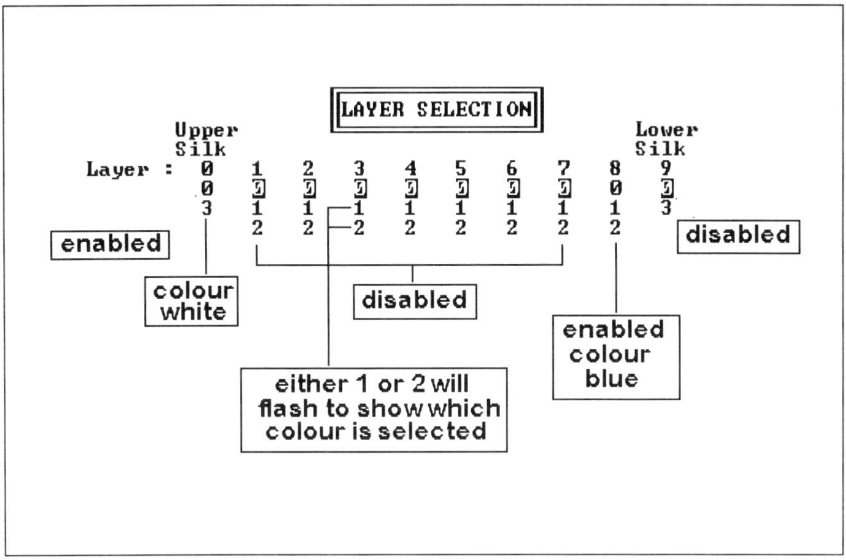

Figure 4.4 Selecting which layers will be used from the User Defaults menu.

3 Make sure that the Grid is on, and set Snap at whole-grid spacing. Set Angle Fix to 45°.

- Always use a whole-grid spacing when you place components. You can switch to half-grid spacing when you lay down tracks, and you may need to switch momentarily to quarter-grid spacing if you need to take a track between IC pins. You should avoid the freehand setting unless its use is absolutely essential (which is very unusual).

4 Load in the transistor shape TO5 (it's an old shape, but easier to work with at this stage). Position it in a convenient place, Figure 4.5.

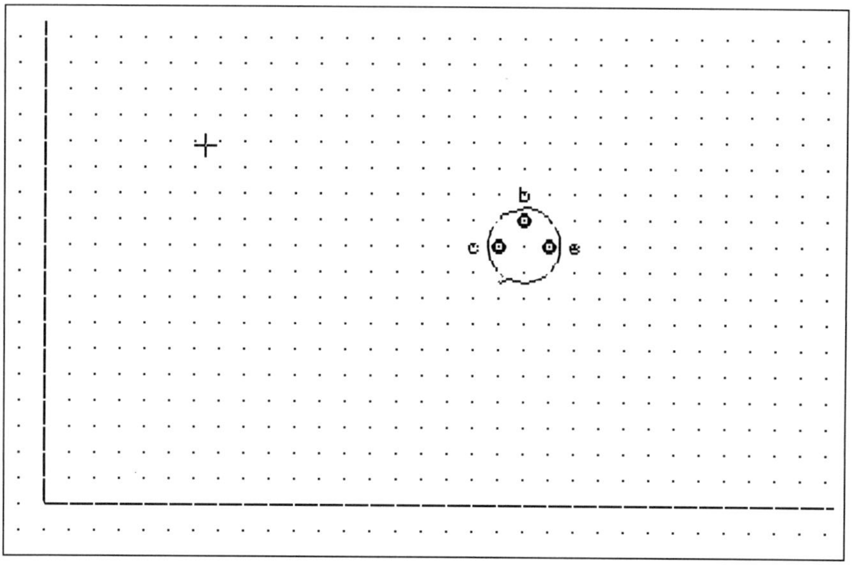

Figure 4.5 The TO5 transistor outline placed on the board.

5 Put in pads for the resistors and capacitors, and label this *silk-screen* – it is advisable to letter the transistor connections to remind yourself that you are working from the component side of the board.

- The normal spacing for pads intended for resistors is at least 0.4". Use fairly large pads, such as the default size illustrated here, unless you have special requirements. Small pads are difficult to etch by DIY methods, and are more costly when the PCB is prepared by way of a bureau.

You can also put in pads for the supply lines, input and output. Figure 4.6 shows the state of the board at this stage. You should save the file at this point.

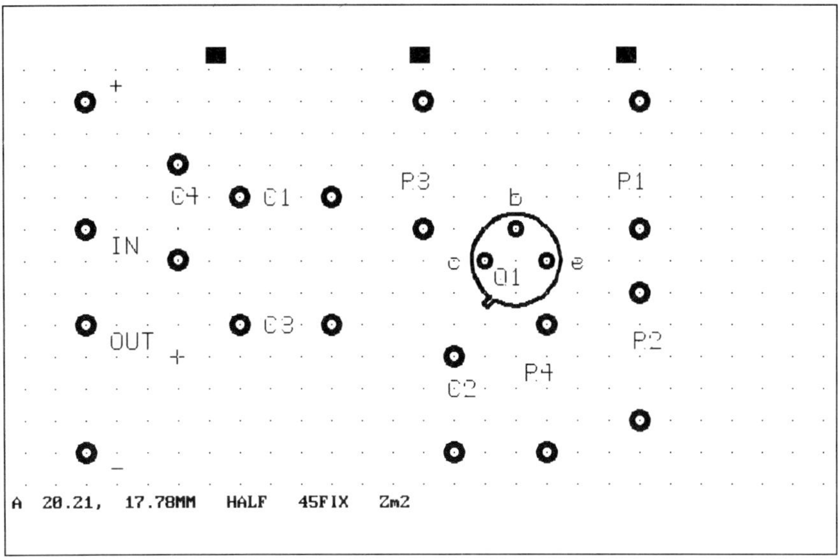

Figure 4.6 The board with the positions marked out for the other components.

This creates the component settings and you can now check this out to ensure that all the components and pads are in place. At present we shall not look at enhancements like using a wide track for earth and positive supply lines, but you should bear these in mind.

- This example is deliberately made simple to point out the steps in using EASY-PC, so that the layout shown here is not claimed to be ideal.

Once the component positions have been settled, you can start laying down tracks. This is done as follows, bearing in mind that layer 8 will be used, not layer 1, in this example.

Before you lay tracks, decide on your track width. If you do not do this, the tracks are likely to be laid using a very narrow track – the line width that you have been using for schematics. You will almost certainly need to fix a new track width before you start laying tracks. If you intend to etch a PCB for yourself using the ferric-chloride method (see Appendix D) then you should use tracks that are around 0.050" wide – this is track width 5 in the default set that is loaded when you install EASY-PC. The minimum track size you should consider even if the PCB is to be produced by a bureau is 0.010". Narrower tracks cannot be reliably produced by DIY work, and will be very costly if produced by a bureau, because the narrowest tracks will need to be cut by laser, making the PCB very expensive to produce.

1 Place the cursor on a pad from which you will lay a track.

2 Press the F2 key for New Track, and press L for Layer – type `8` for the layer. If you do not do this, the layer 1 will be used. Press `W` for Width and key in `5` (for 0.052" width) unless you intend to use 0.010" tracks (key in `3`).

- There is nothing wrong about using layer 1, the default, for connections that will be made on the copper side of a single-sided board, but, as noted earlier, this gives you an opportunity to practice working with other layers.

3 Join your first track, clicking the left-hand mouse button to fix the track. The track should appear in blue.

4 Place the cursor where you want the next track to start and use F2 again. Click where the track is to end.

5 Continue until all of the tracks are drawn in – remember that you are viewing this from the top (component) side of the board. Save the file again when you have finished.

- If you make an error, use the F1 key (Edit Track) to delete a track section. Be careful to ensure that each track starts from, ends or passes through the centre of a pad – with a small snap setting it can be only too easy to create tracks that do not quite reach their pads.
- Try to make all of your joins at 90°. If one line joins another at 45° or less, the joint will be a poor one, because the etching process will eat into the narrow angle.

The board should now look something like the illustration of Figure 4.7. All of the component positions and the tracks are now in place and we are looking at both front and rear as if the board were transparent.

We now need to make a border and markings that will show the limits of the board and identify which way round it is. Figure 4.8 shows this border drawn (as a silk-screen, and so in white), and pads placed at the corners as locating points. To draw this, ensure that you are not in any editing mode (press the ESC key or the right-hand mouse button) and place the cursor where you want the border to start. Press the F2 key (New Track) and select layer 0 using L and 0 keys, then width 2 by using W and 2 keys. Draw the border just as you drew the PCB tracks. Place a pad at each corner, because this can be useful for alignment purposes.

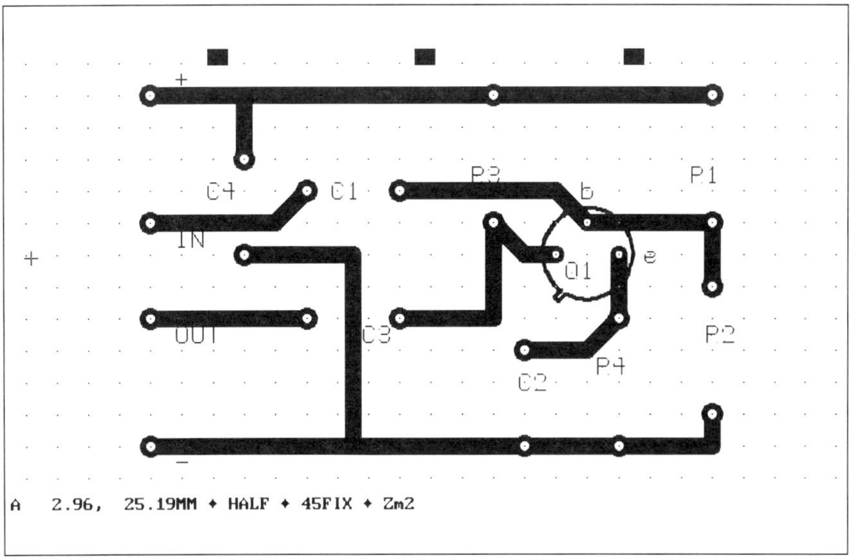

Figure 4.7 The connections for the circuit laid down. In a simple circuit like this there are no problems of crossing tracks.

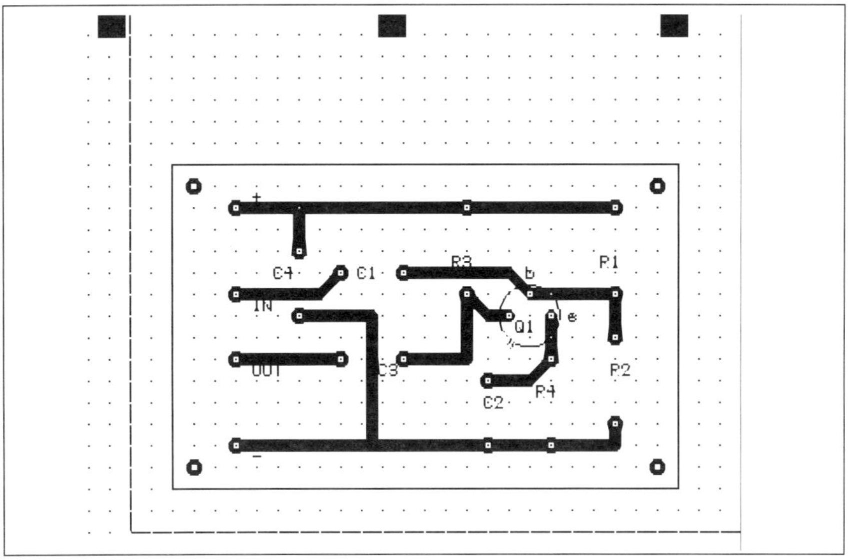

Figure 4.8 The schematic with a border drawn round it and pads placed at the corners to define the final board size and shape.

- Remember that these images will be in colour when you are using EASY-PC for yourself.

Now you can print this, printing all layers to see both silk-screen and track sides together. This can be a useful print to retain, because it is much easier to check than a set of separated views of silk-screen and track layers. Finally, you will want to print both the silk-screen and the track views, with the track side flipped so that you can use it to make the etching mask.

1. If you are printing on a dot-matrix machine, use *User Defaults* (F2) to turn off all layers except layer 8, the track layer in this example. If you are using a laser or H-P compatible ink-jet printer you can select your layers individually with EASYLASE: see Chapter 3.
2. Select your track layer diagram with a block, place the cursor to one side, and use the *Flip* command of the *Block* menu.
3. Save this version of your layout using a different filename, preferably one that reminds you that layer 8 is flipped over to look at the solder side view.
4. You can now print this view, Figure 4.9. You can use the printing utility to print each layer on separate pages if you want, so that you have both a silk-screen view and a soldering view – see the EASY-PC manual for details.

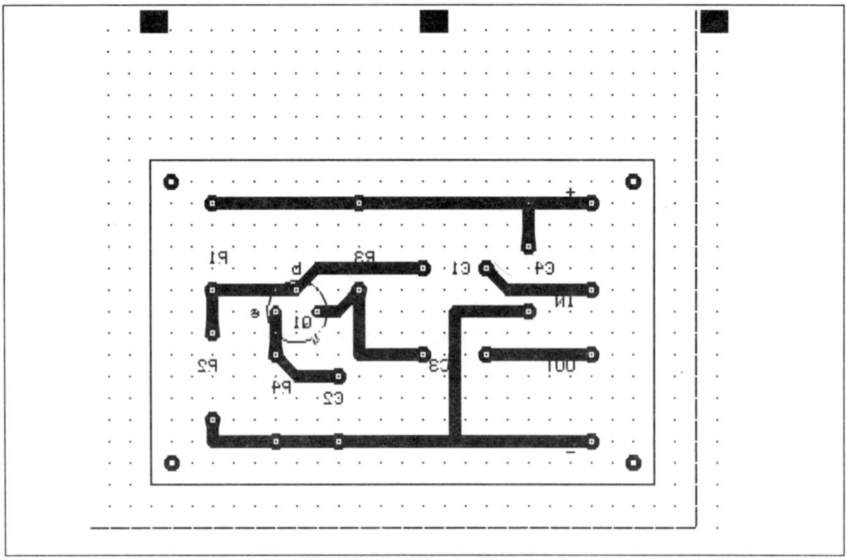

Figure 4.9 The underside copper strip view as printed. Remember that the normal view of this layer in EASY-PC is from the front silk-screen layer.

Unless you have considerable faith in your understanding of EASY-PC you should work through this simple example for yourself, because if you encounter any difficulties they should be relatively easy to sort out, something that cannot be said for more elaborate layouts. The next step is to look at a digital circuit layout, still using a single-sided board.

A four-bit ripple counter

The use of discrete component circuits, though by no means dead, is only a very small part of PCB requirements nowadays. When you start to work with ICs the layout of a PCB can become very frustrating unless you happen to be one of these people who can see a way round a maze in one glance. In particular, you are likely to find that in the normal way of working, a lot of trial and error is needed, and a vast amount of paper can be consumed trying to make a set of sensible connections, particularly on one-sided boards. It can be very tempting to assume that IC layouts can be done only on double-sided boards but single-sided boards were in general use for all IC layouts before double-sided boards became generally available.

- This does not mean that every connection must be made by copper strip. You may need to use a wire link if the only other options would involve long and tortuous paths or the need to use very narrow tracks between IC pins.

The example we shall look at as a gentle introduction to using EASY-PC with IC layouts is a four-stage ripple counter using a couple of J-K flip-flops. The flip-flops are 7473 types, with a clear terminal but no preset, leaving out one set of connections that might cause bother. As always we need to start with a schematic, and in this case it has been labelled with pin numbers as well as with connections. Note the important supply pins – it is very easy to forget them in a diagram like this.

1. Start EASY-PC running, and from the main menu select Design Schematic, item 2.
2. Start with the grid on, and use zoom level 2. Make sure that you are using whole-grid snap (key M). Use New Symbol to place one component symbol, filename JK+C, in place.
3. Use the cursor and the R key to put in three more copies, and unzoom by one level so that you can see what you have done, Figure 4.10.
4. Now draw in the circuit lines, using 90-degree fix and half-grid snap.
5. Put in the text for the J-K labelling, and the important pin numbers, Figure 4.11.

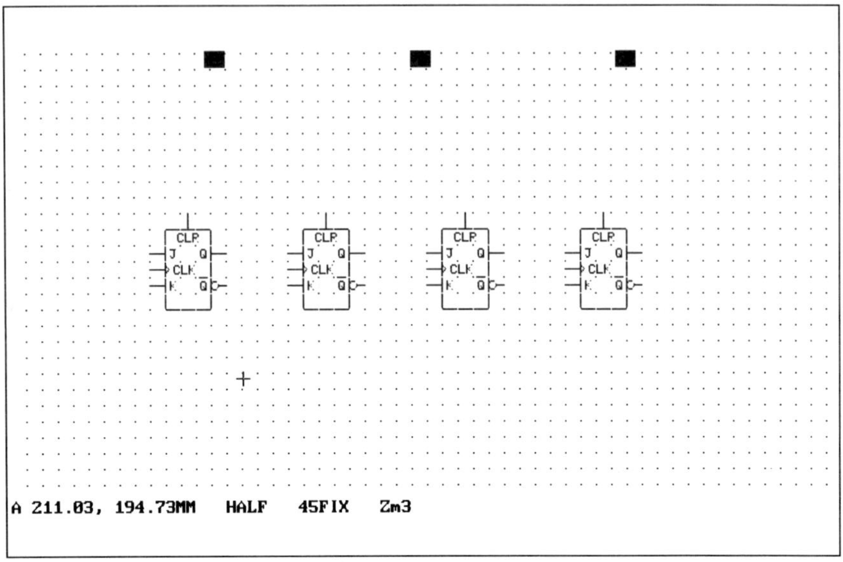

Figure 4.10 A set of IC symbols placed into a schematic.

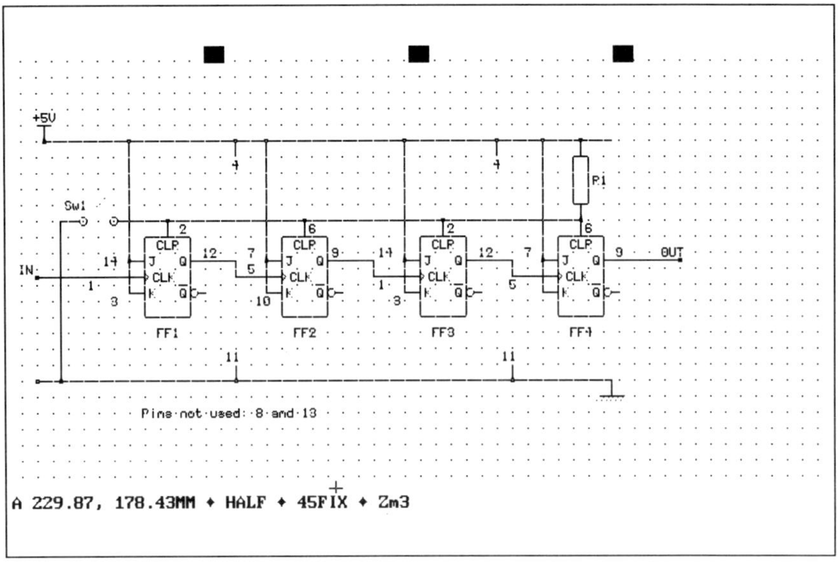

Figure 4.11 Starting the counter schematic, with pin numbers added along with a note of unused pins and the pins used for the power supply positive and negative.

6 For a circuit like this, which is normally wide but not tall, it is an advantage to print landscape. When you have saved the schematic, rotate it through 90° and save this under another filename. Use this rotated version for printing, as it allows you to print at as large a scale as your printer and paper size can achieve. This makes it easier to see the numbering and lettering.

- You would not normally put in pin numbers for a schematic, but when you intend to prepare a layout from that schematic it makes the task much easier if pin numbers can be seen. It is also useful to note pins that are not used in the circuit and pins that are not connected, if any.

The next step is some preliminary work. There is no simple correspondence between a schematic and a layout diagram, where ICs are concerned, and it is unusual to find an IC layout that allows you to make connections that look anything like the lines of a schematic. This becomes particularly difficult when a package contains more than one IC, as is normal for the smaller-scale items such as gates and flip-flops.

- If you are experienced in PCB layout, then you can ignore the following, but for the inexperienced user, or for anyone who has laid out circuits only on stripboard, the advice will be useful.

1 Start with a new blank PCB layout, item 1 from the EASY-PC menu.

2 Zoom this up to level 2 and use *New Symbol* to put in the layout for one copy of the pin set for the main IC that will be used, in this example the 14-pin set for the 7473. This is done by placing the cursor where you want the origin of the symbol to appear, pressing the F8 (New Symbol) key, and typing 14 (for a 14-pin IC) when you are prompted by the message at the foot of the screen. You will then be asked which symbol library to use, with the default of the standard library showing. Press the RETURN key (or left-hand mouse button) to confirm this library and load the shape.

3 Label this with the pin connections of J, K, Ck, Clr, Q and the supply + and – pins. Add the numbers for pin 1, pin 7 and pin 14. Remember that for this IC there will be two pins for each function other than the supply + and –. Remember that this text will be on layer 0, the silk-screen layer.

- You can place text on a copper layer so that the text appears in etched copper. This would be used for a reference number, date of design, or other data which you would want to have in more permanent form. Such copper lettering must be kept well away from tracks or components.

4 The result is shown in Figure 4.12.

- If you have to shift the position of this symbol and you find traces of the previous position remaining on screen, place the cursor in the middle of the drawing and use the P key (Pan). This will always re-draw the screen.
- If you want to print this diagram, you can do so in the usual way. It is very much quicker, if you are running EASY-PC from Windows, to press the Print Screen key, and switch to Windows Paintbrush. Use Edit-Paste to put the image into Paintbrush and then print it from Paintbrush. This is very much faster for an image like this than printing from EASY-PC. You can also use Paintbrush to invert colours, which will make the image easier to see.

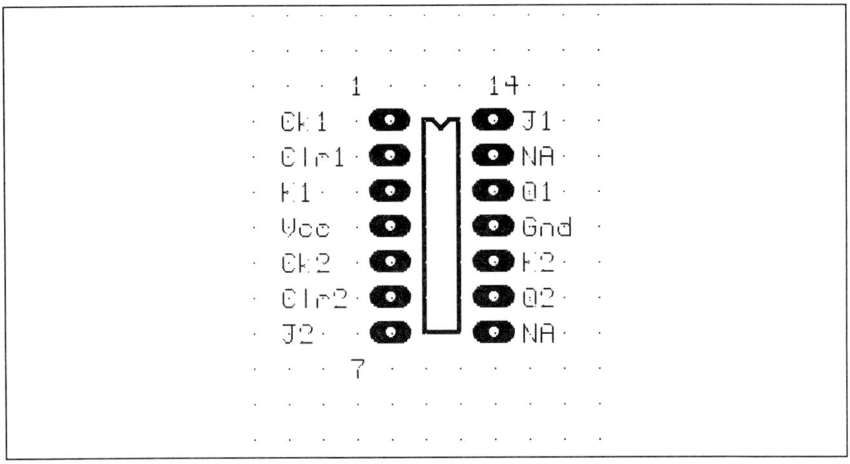

Figure 4.12 A zoomed section of the PCB layout when one of the flip-flop chips has been put in and labelled.

The purpose of doing this first step is to see how best to lay out the component. It is very tempting, when you have little experience of PCB layout, to place the IC with pin 1 upwards on the PCB, and with inputs coming from the left, and outputs going to the right, with the + supply line above the IC and the earth line below.

In almost every case, this leads to impossible layouts, with lines crossing each other all over the place. We need to find what way of positioning the IC relative to the interconnections will be easiest, and there is no fixed rule that can be applied. The following steps, however, will guide you.

1 Look for connections to bus lines such as the Vcc and Gnd lines. By convention, such lines will run horizontally on the top copper layer (layer 1) and vertically on the bottom copper layer (layer 8).

2 Look at internal cross-connections, where pins of the IC are to be connected without affecting any other IC.

3 In this example, J1 and K1 are connected, as are J2 and K2, Q1 and Ck2, Clr1 and Clr2.

4 There will also be external connections for J and K and the Clr pins, and the supply terminals. Indicate these, but do not at this stage try to work out where they will go, Figure 4.13. The lines that go to buses are drawn in width 5 (0.052") and the lines that pass between pins in width 3 (0.010").

- Note that the pins labelled as NA are not used (they are the Qbar or notQ pins), but they are internally connected and must *not* have connections made to them. You can label unconnected pins as NC.

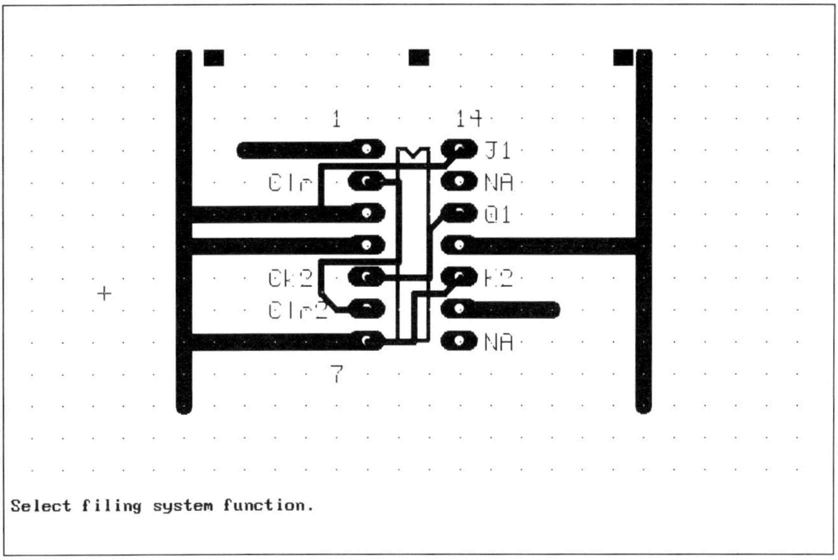

Figure 4.13 Placing connecting tracks to one flip-flop unit, concentrating on connections between pins within the same chip.

5 In this example, some changes have been made from the initial layout, because the connection of the + pin and the K1 pin to the supply bus prevents a straightforward connection being made from Clr2 to Clr1, pins 6 to 2.

6 What is more important is that when you look at this in conjunction with the schematic, you will see that the layout will be much easier if the ICs are laid sideways with the pin-1 end facing left, Figure 4.14. This allows an uninterrupted bus to be used for each supply line, and it also makes it easy to arrange connections between the Clr pins of another IC.

- You may need to alter the default Snap setting of half-grid spacing to quarter-grid spacing in order to make some of these tracks between pins of the ICs. Work with half-spacing as much as possible.

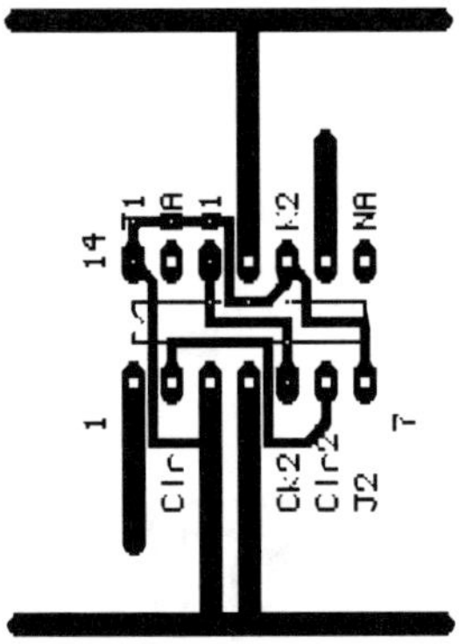

Figure 4.14 A portion of the view of the layout with the chips rotated to be end-on to each other. The silk-screen layer is also shown so that you can see the IC outline.

7 Looking at this, you can see other possibilities. It might be neater, for example, to connect J2K2 to J1, so that there was no need to connect J2 directly to the supply line. This leaves the Clr2 pin free for a connection across the Clr1 pin of the next IC, Figure 4.14. This view has been rotated, using a block anticlockwise step.

- This view has been captured with only layer 1 visible (using User Defaults to switch off the other layers). Note that text is still visible. When you place text on a PCB you must remember to switch to layer 0 so that the text is on the silk-screen layer. The only exception is to identifying text that should be in copper. For example, you should always identify the side (Side 1, Side 2) of the board in this way.

This preliminary work may look tedious, but it saves an immense amount of time in the long run, because it is possible to spend (literally) hours on a layout that seems quite impossible because you are simply looking at it in the wrong way.

- Always cross off a connecting line on the schematic each time you complete a connecting track. This makes certain that you do not forget connections.

You now need to check this carefully to be certain that you have not programmed any impossible routes, nor made any mistakes over pin numbers or uses. The next set of steps will end in connecting up the second IC.

1 First delete the text from this layout. Call up the User Defaults (↑F2), and look at the section marked *BLOCK SUBMENU*. You want to use a block delete that will affect only text, so you need to turn off all but text in this set. Press the keys H, I and K so as to change ON to OFF for each of these items, Figure 4.15.

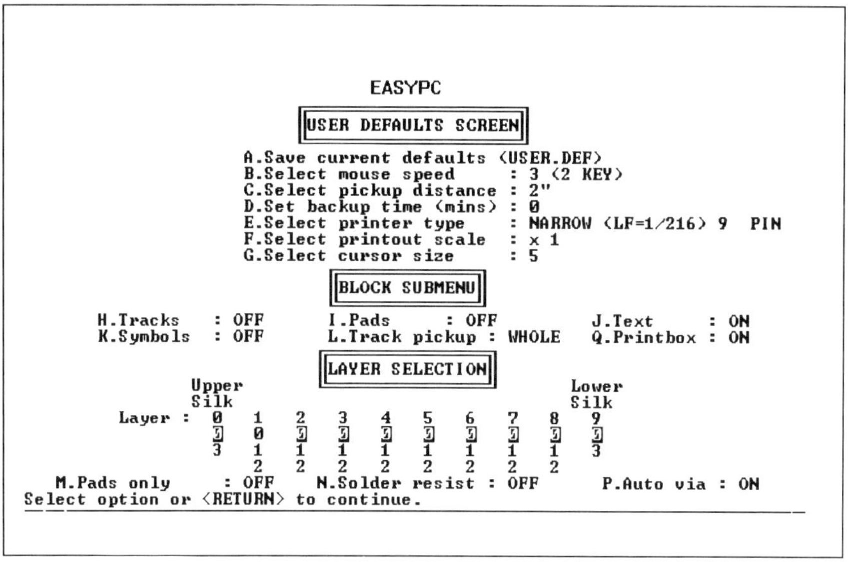

Figure 4.15 Preparing to make a copy of the IC text without copying the tracks, pads or symbols.

2 You can add pin 1 and pin 8 identification numbers to this after restoring the defaults.

3 Restore the Block action to all objects, and use a Block Copy action to make another IC with its connections, and join up the sets of leads (Figure 4.16). This exposes a problem with the connections of pins 6 and 2 to the corresponding pins of the next IC.

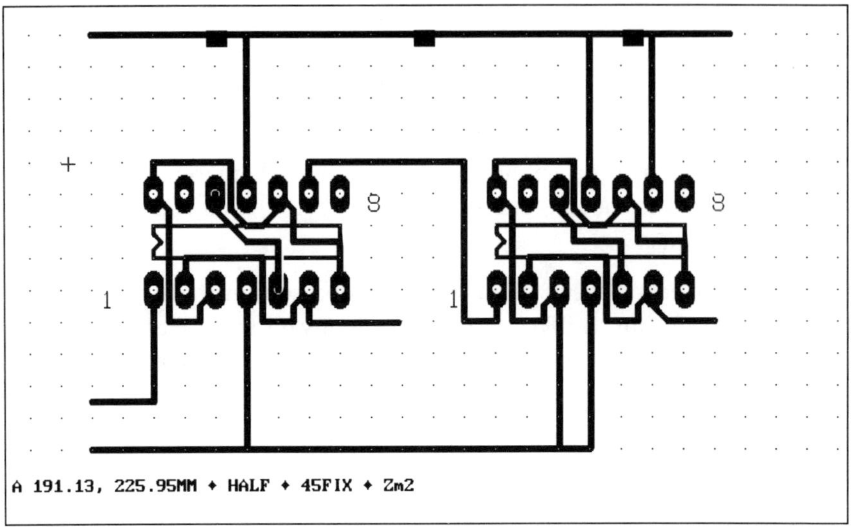

Figure 4.16 The second IC in place, using the same connections within the chip.

4 This is solved by some re-routing, Figure 4.17. The main supply line connection is now from pin 7 of the second IC, and the other connections are altered to suit.

- You could not get away with this re-arrangement if there were more ICs to follow, so that this fudge indicates that more time should have been spent on looking at connections when a single IC was being viewed.

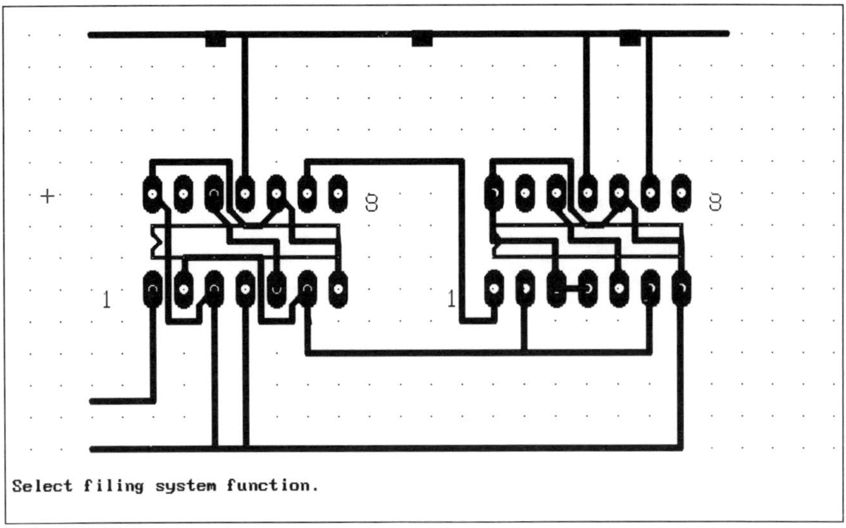

Figure 4.17 Re-routing lines around the second IC.

The pads for the reset switch can now be added, and the time has come to give some thought to tidying up, in particular the size of the PCB. The earth and supply lines are too far from the ICs, and this can be sorted out by using another block shift. This time we want to throw a block around the positive rail, and move it closer to the ICs. We want to do this in such a way that the existing connections remain but will shrink to size.

1. Use the User Defaults menu again, and switch off Pads, Text and Symbols. Use the L key to change Track Pickup from *WHOLE* to *PARTIAL*. This means that a track that is completely encircled will be moved but will retain its connections outside the block.
2. This allows you to move the main ground track further in. You can also move the others closer, tidying up the connections.
3. The switch pads also can be moved to allow for closing up the gaps.
4. Finally, we can use wider tracks for the supply lines. Place the cursor on the track, press F1 and use W to get the width action. Try typing 4 as the width number and press RETURN to see the track widen. Figure 4.18 shows the results on a zoomed view of part of the layout.

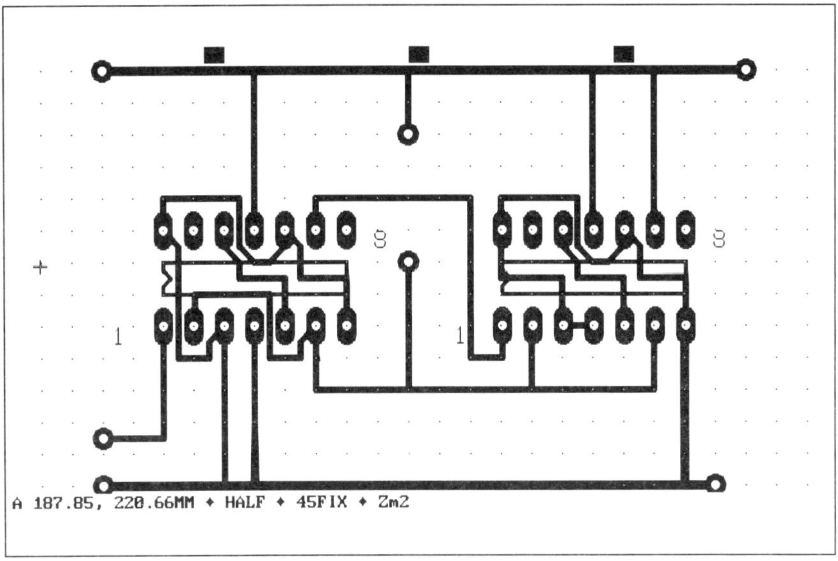

Figure 4.18 The final layout for the counter circuit.

- If your supply tracks consist of several segments or if you have crossing lines that share a node position you might find it difficult to locate the segment you want. Where crossing lines share a node position, you will find that pressing the F1 key will toggle you between the nodes that exist at that point. If you have a lot of short segments near to each other, press the F1 key with the cursor unequivocally over one node in a segment (you might need to use the N or B keys to get to the segment you want).

- Always check, using the H, N, B and E keys, which line the cursor is on before you use any actions that would change tracks.

Odds and ends

Despite the heading, these are of considerable importance.

- You can see what attributes (width, layer etc.) exist for any entity (line, pad, symbol or text) if you put the cursor over the origin of the entity, press the appropriate editing key (F1, F3, F5, F7) for the entity you want information on, and press the large + key on the number keypad. The display, Figure 4.19, is visible only while the key is depressed. If you are running under Windows, ignore any beeping sound you hear while the key is depressed.
- When you edit any entity, EASY-PC will retain the attributes of the edited entity until you change them. For example, if you edit a line with width 5, any line you draw from then on will be of this width. This makes it possible to keep template lines and pads, for example. You can place, well away from your drawing, a set of the different lines and pads that you intend to use. To switch from one type to another, simply place the cursor over a line or pad example and press the appropriate editing key. This makes it much easier to enforce uniformity.

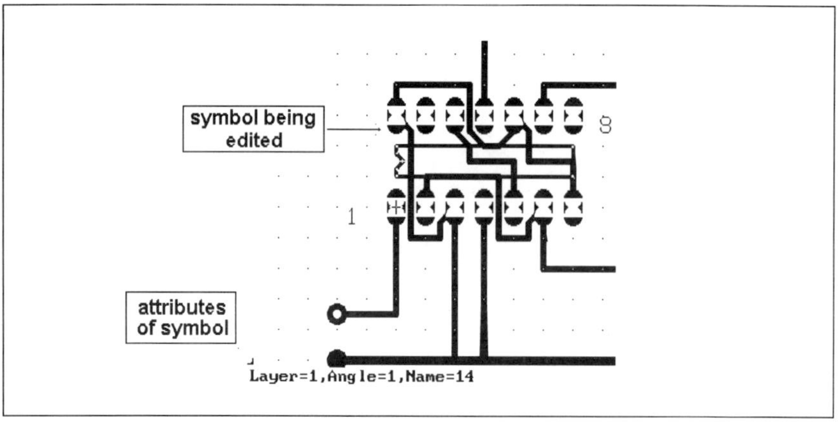

Figure 4.19 The IC symbol being edited.

5 More complex layouts

Editing tracks

NOTE: Examples in this chapter show thin tracks because this makes it easier to follow the routes. You would normally aim at using a minimum track width of 0.010".

The advice in Chapter 4 concerning planning layouts becomes increasingly important when you start to work with more complex layouts using ICs of greater scale of integration. In general, circuits that use only the older varieties of TTL chips can always be constructed on single-sided boards, but later types of MSI TTL chips and virtually all chips associated with a microprocessor will require a double-sided board layout. Even the simpler layouts, however, can often benefit from using a two-sided board because this can usually avoid some of the tortuous routes that are needed for a single-sided layout. A good rule is that if your single-sided layout would demand the use of more than one track passing between pins of an IC then it's time to look at a two-sided board or to put in pads for wire links.

- If you use the increasingly-popular silk-screen method of depositing resist you will find it difficult enough to get one track between IC pins – the secret is to use rather smaller pads.

The amount of planning that you use for more elaborate layouts, however, should be confined to planning the best position and orientation for the symbols. If you were to plan each track there would, after all, be little point in using EASY-PC (except to make a neater drawing). We need now to take a look at one of the advantages of using EASY-PC, the ability to edit tracks.

Figure 5.1 shows a rather more elaborate digital circuit, a four-stage synchronous counter. This has been drawn using the schematic option of EASY-PC, and the illustration is taken directly from the screen. The hardware consists of two flip-flops, two gates and two resistors along with a push-button switch.

Start the layout now by checking your settings. Return to the main menu and take the PCB layout option. The Grid and Snap should be on, using quarter-grid snap, and 45° angle enabled.

Figure 5.2 shows the IC components positioned on a board, with no attempt as yet to prepare tracks. The resistors and the switch have not been positioned yet because we need to see how the tracks will lie before putting in pads for these components.

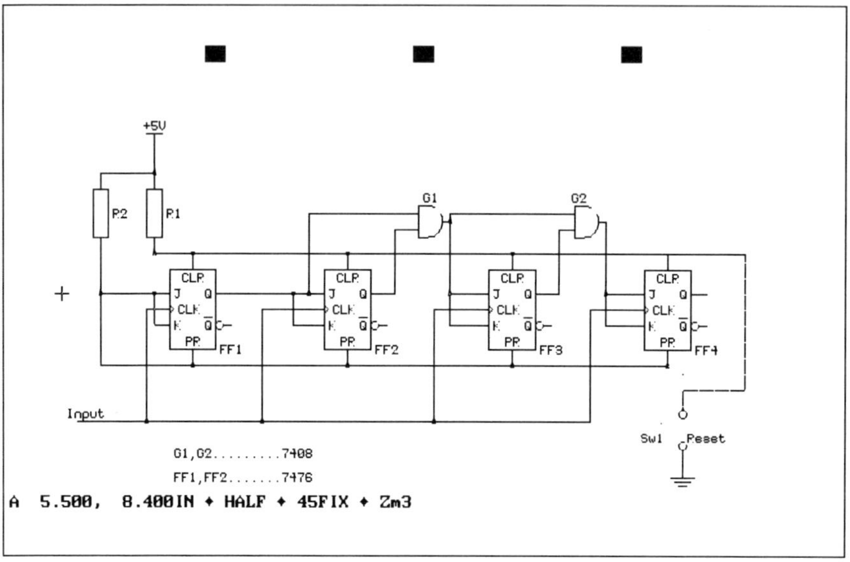

Figure 5.1 A circuit for a four-stage synchronous counter drawn using the Schematic option of EASY-PC.

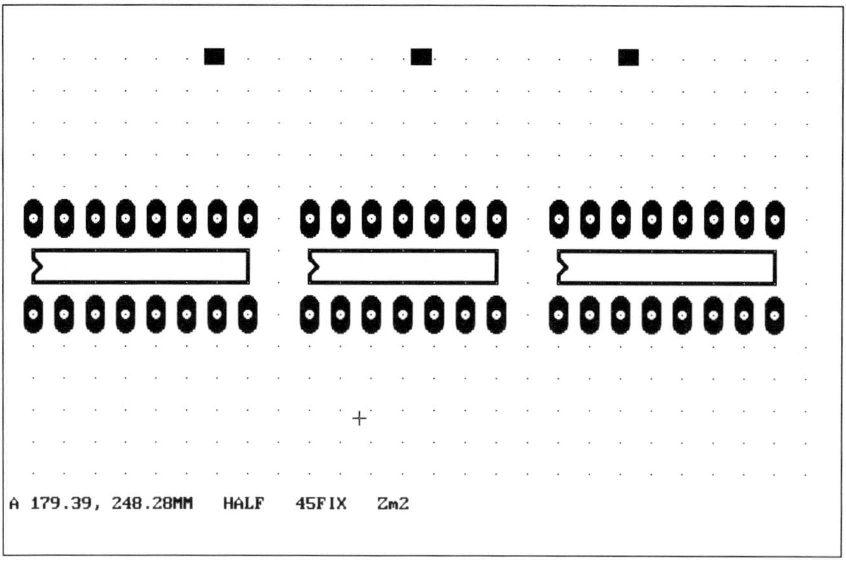

Figure 5.2 The active components laid out on the board, with the gate chip between the flip-flops. The resistors and switch can be left until later.

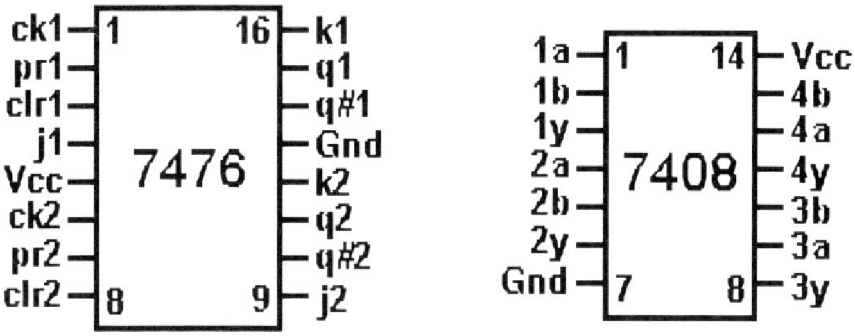

Figure 5.3 The pinout connections for the two chip types. It is useful to have a large-scale copy of this when you are laying down PCB tracks.

Before we start to lay tracks we need to have pinout information, and Figure 5.3 is a pinout diagram for the two types of ICs. The pinout for the 7476 indicates that once again it will be an advantage to position these chips in line, as was illustrated in the previous chapter. You should now write in the pin numbers on the schematic unless you did so when the diagram was drawn.

- These pin assignments should not be printed on the PCB – either put pin numbers on the schematic or, as shown here, print a separate pinout diagram.

Note that you can arrange the chip shapes vertically or horizontally; it's a matter of convenience only. In this example they have been arranged horizontally and packed rather closely so that the three just fit on the screen at zoom level 2. Normally, you would want to leave the chips further apart until you had completed the tracks and then use a block move to tighten the layout up again.

We start by putting in tracks but this time the tracks are drawn from point to point with no regard to crossing, because we want to look at the methods that are used to untangle the rat's nest layout. Quite a number of tracks can be laid without conflicts, but in any but the simplest of layouts there is bound to be some crossing, and we can look at these to determine whether the conflicts can be resolved using a single-sided layout or if a double-sided board will be needed.

- The tracks are drawn using narrow lines in the following illustrations to make it easier to see lines passing between pins. You would normally use at least 0.015" tracks for single-sided boards.

The layout now looks as in Figure 5.4, and there are a lot of crossing tracks. The gate chip looks as if its connections would be better arranged if the chip were rotated by 90°, and though we can do this with a block action, maintaining connections, it is easier in this example to delete the connections and then turn the shape 90° anticlockwise with a block action.

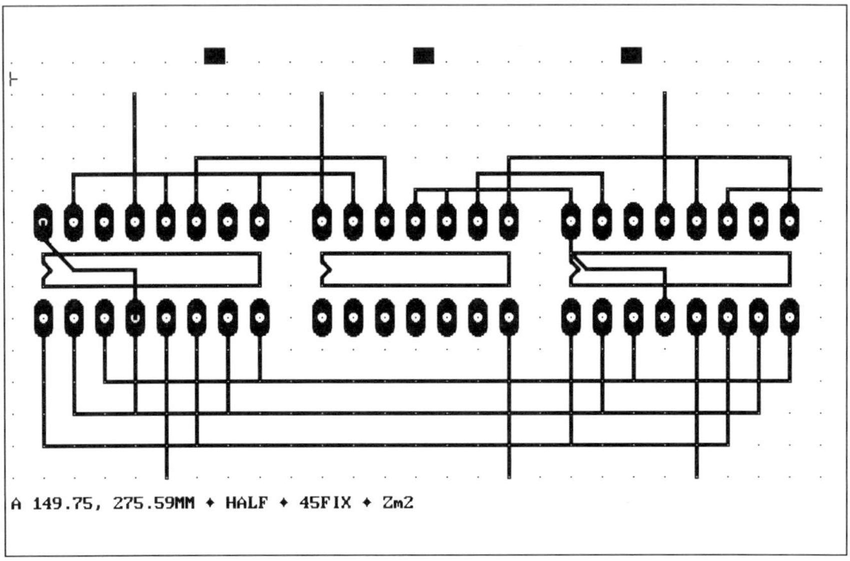

Figure 5.4 A first attempt at putting in tracks. Looking at this suggests that the gate chip might be easier to connect if it were turned.

- Note that I said *in this example*. For anything more elaborate, you would by this time have reached a stage where you would not want to break any connections – in the language of PCB layouts, you need to maintain connectivity. When you have been methodically crossing off each connection on the schematic as it was made on the layout, breaking connections is the last thing you want, and if you shift or rotate a chip you should use the block method with Partial Track Pickup so that connectivity is not compromised.
- Another point is that rotating a chip is all very well if you are making a PCB using DIY methods and for your own use only. If you are designing a PCB for commercial production, all chips must point the same way. A design in which chips are likely to point in different directions, particularly if some chips are turned through 180°, is asking for a high reject rate on inspection.

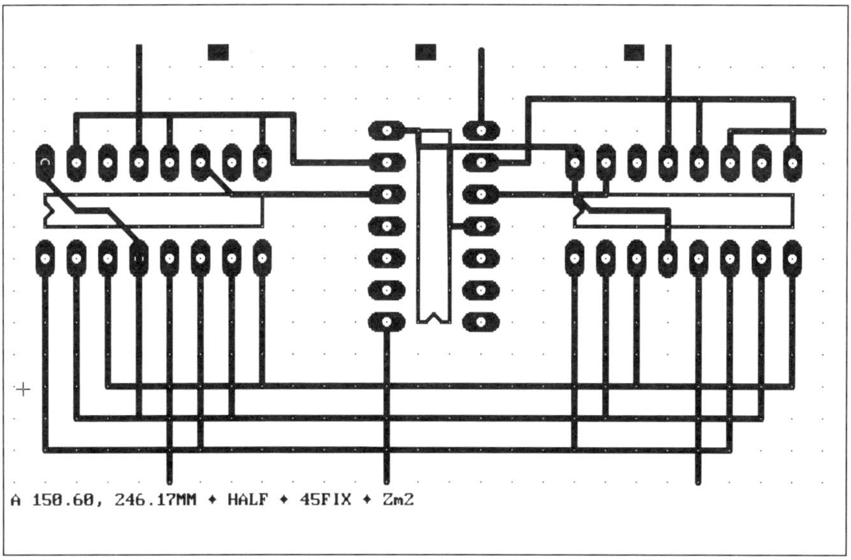

Figure 5.5 The appearance of the tracks with the gate chip rotated.

This has been done in Figure 5.5, using the same zoom. The gate chip has been turned and connections have been selected which look reasonably easy to use. The nest of connections under the chips has been pulled down by using User Defaults to alter Track Pickup from Whole to Partial. This allows the tracks to be selected and pulled down without breaking connections – the track lines are simply extended. One point to note when you use Partial Track Pickup in this way is that tracks can end up at angles other than 90° or 45° if you move the block in a diagonal direction. This will require track editing to restore the normal track directions, but the time spent in editing the tracks is well spent if it avoids breaking connections – connectivity is more important than any other factor, because re-checking a layout from a schematic requires a lot of time and is prone to errors.

With the layout now looking as in Figure 5.5 it's time to edit tracks so as to avoid cross-overs if this is going to be possible. Once again, practice will help you with this, and some users find it much easier than others. Someone who lays out one PCB per month is likely to find these actions more difficult than another person who does two per day.

Figure 5.6 shows the first stage in uncrossing tracks. The connections around the gates have been re-routed first, followed by the cross-over of the earth lead of the first flip-flop and the signal connections. The general method is:

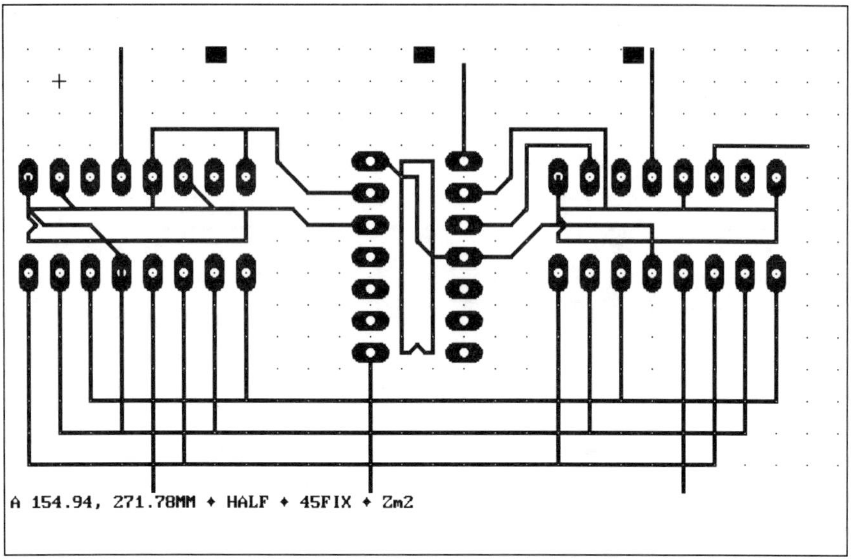

Figure 5.6 Re-routing started, with several tracks in their correct places.

1. Place the cursor on the track that is to be re-routed and press the F1 key to edit the track.
2. Use the B and N keys until the required section of track changes colour (from red to white in this example).
   - You can reverse the Begin and End nodes by using the Invert command (↑I) during track editing. This can make track editing easier if you need to extend a track.
3. Use the D key to delete this node and, with rubber-banding on, move the cursor so that the line follows the direction that you want. If the new route requires more sections, click the mouse button and repeat the action with any new segment that is still faulty.

This also needs practice, because the incorrect track will still be visible until you click the mouse, and this can be confusing. In addition, if you are re-routing in a tortuous path to avoid a cross-over, you may need to delete and rubber-band several times.

The illustration has been printed with layer 0 visible. You can, however, switch to User Defaults with ↑F2 and press the zero key. This removes the symbols and makes it much easier to see tracks that are running down the outline of the symbol. The only snag with this is that it makes it more difficult to identify pin 1.

By this stage, the layout is beginning to clear, and the three leads to earth for the ICs are now clear of any signal lines. The next stage is to deal with the interconnections that on this pattern are positioned below the chips.

As the main conflict is caused by the positive supply leads from pin 5 of the flip-flops the procedure is to re-route these lines, either picking a route if you can see one, or leaving the pins unconnected until the signal paths have been resolved.

1 Start with the lines that connect to pins 1 and 6 on the first flip-flop and route these outside the pin lines.

2 Now pins 16, 2, 4 and 7 can all be linked through pin gaps and with leads inside the pin rows.

3 Pins 3 and 8 can be linked and the Vcc line on pin 5 can be taken out between pins 1 and 2. Figure 5.7 shows the results of all this.

   - Note that if you are unhappy with taking relatively wide tracks between the pads of an IC, you can neck the track where it passes between the pads. Remember that you cannot alter the pad sizes on a symbol except by editing the symbol and preferably placing it in a different library. To make a necked crossing, Figure 5.8, take the track between the pins with a node at each side and complete the track in the normal way. Now edit the track and use the N key so that the segment between the pins is highlighted. You can now

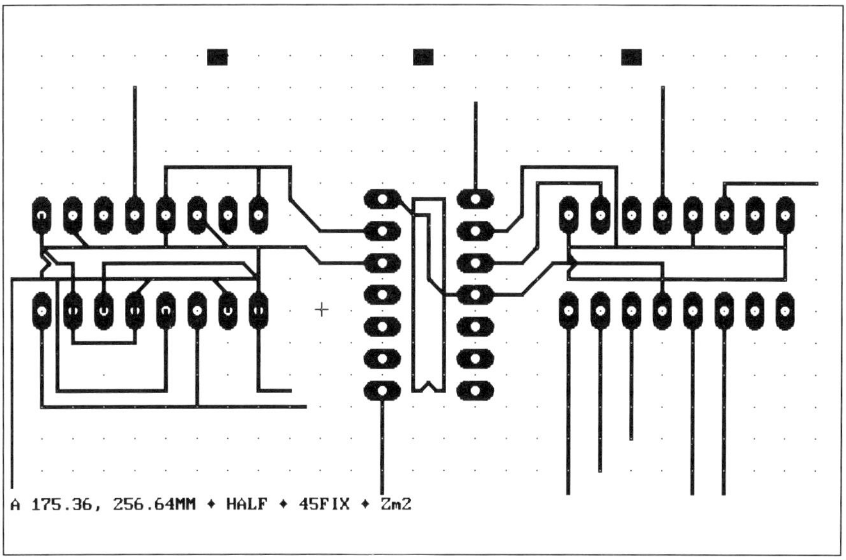

Figure 5.7 The first flip-flop is now free of crossed tracks.

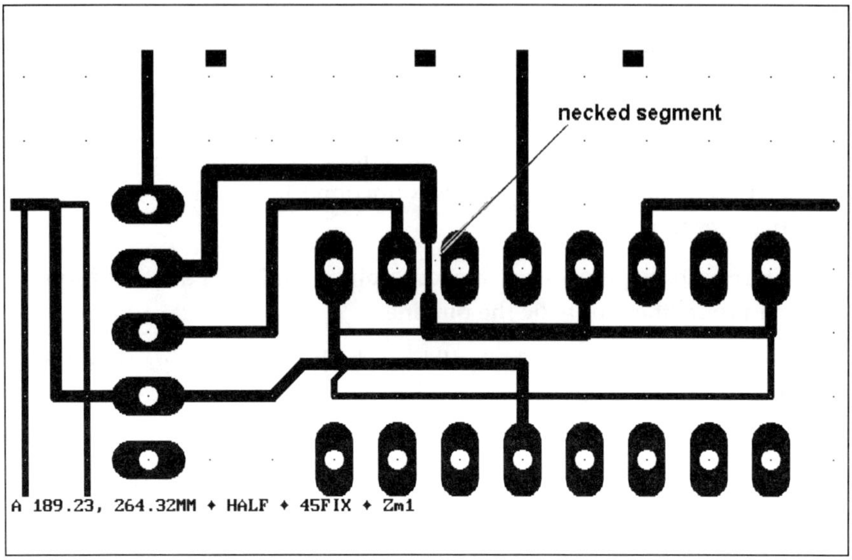

Figure 5.8 How a segment of track can be necked between pins by using the Segment Width editing command.

use the Segment Width command (↑W key) to specify a smaller width number for this segment only. Remember that if you use the All Track Width (W key) you will alter the whole track, not just this one segment.

The result clears all crossings from the first flip-flop, but the interconnections are still left dangling.

The second flip-flop is not so easy to deal with, but all the signal lines can be joined, Figure 5.9. This leaves only the supply lines, the resistors and the switch, and because the Vcc and Gnd connecting pins of the flip-flops are connected to the opposite sides relative to those of the gate, we shall need to use wire links at this stage.

We can also add the other pads for connections or the resistors and the switch. Figure 5.10 shows these and the links – the labelling for the switch and the resistors is not part of the PCB layout. You would now review track thicknesses, making sure that 0.015" tracks were used for signals and pin connections, with wider track for bus lines. Finally, you would want to ensure that you had pads for input and output and you would want to add text to identify these points and also the Vcc and Gnd lines.

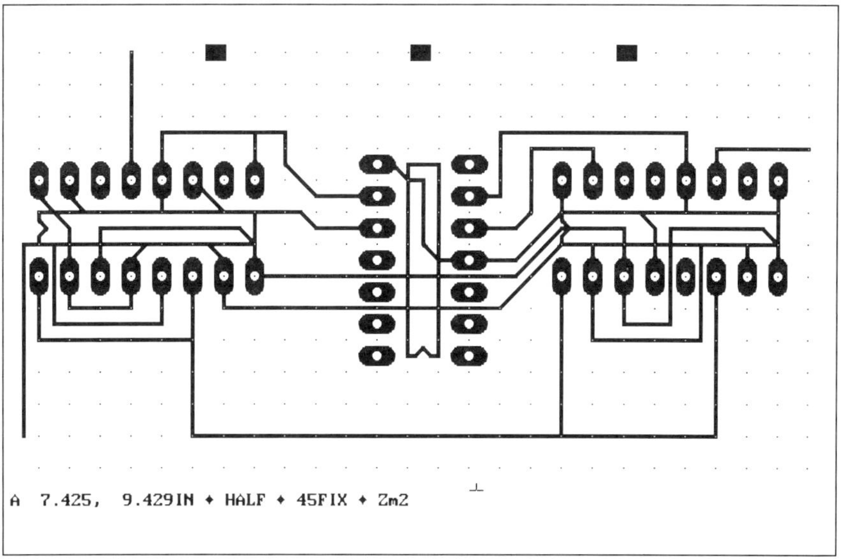

Figure 5.9 On the second flip-flop, the tracks have been laid so that two wire links will be needed to preserve connectivity.

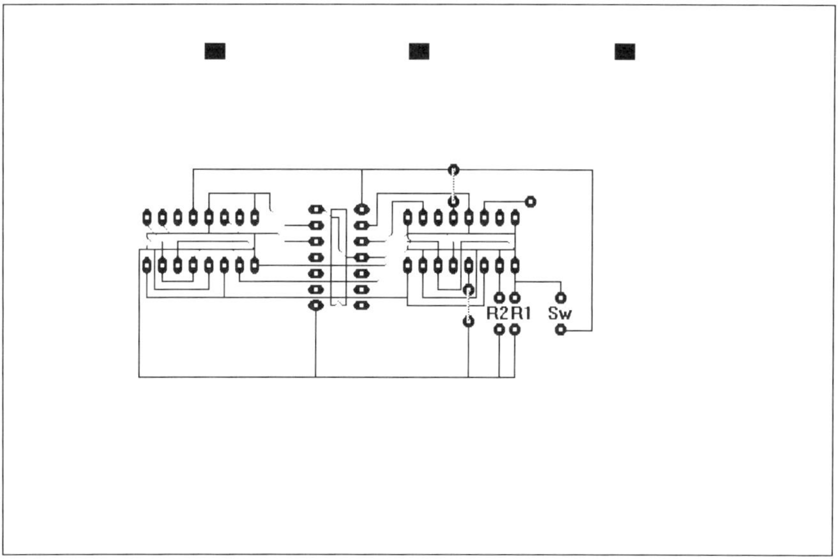

Figure 5.10 The layout with pads added for the passive components, and a border drawn.

Using Track Delete and Zap

If you have worked your way through the example above, you are likely to have experienced some frustration using the track editing actions of Delete and Zap. This is a matter of experience, and once you grasp the principles of track creation and editing you will find it much easier. These basic points were noted earlier in Chapter 3, but now that you have had some more experience we need to reiterate them.

1 A track starts at a home node and ends at an end node.

2 There may be intermediate nodes between home and end – wherever you click the left-hand mouse button will put in a node, and a node will also be automatically put in where there is a change of direction such as you will see when working with 45FIX.

3 When you edit a track, you must locate the cursor by using the (H)ome, (B)ack, (N)ext and (E)nd keys.

4 You start editing by placing the cursor on a node of the track (or as near as possible to a node) and using the F1 key.

- Note that you sometimes pick the wrong track in this way. The track that will be edited is the one which has a node nearest to where you positioned the cursor, and this might not be the track you want to edit. To be certain, put the cursor at the start or the end of the track you want to edit, or any node.

The Delete and Zap actions are quite different. The Zap action (↑Z) will remove all of a line from Home to End – but it will not affect any other line that has been drawn to meet that one. The Delete action will remove a node and EASY-PC will re-join the remaining nodes. This is best done with rubber-banding switched on (key X).

Figure 5.11 shows an example of two lines. The horizontal line has been drawn too long, and the vertical line has been drawn from where the horizontal line should have ended.

In this case, when you use F1 and H the whole of the horizontal line is highlighted. When you use (E)nd followed by the D key you can then move the cursor to the point labelled as End 2 in the diagram. Press the left-hand mouse button (or the RETURN key) to fix this in place.

Another solution is to use the Invert action (↑I) which interchanges Home and End. You can now use the D key (b), and with rubber-banding on you will see the end of the horizontal line rubber-band so that you can move it to the start of the vertical line, press RETURN (or left-hand mouse button)

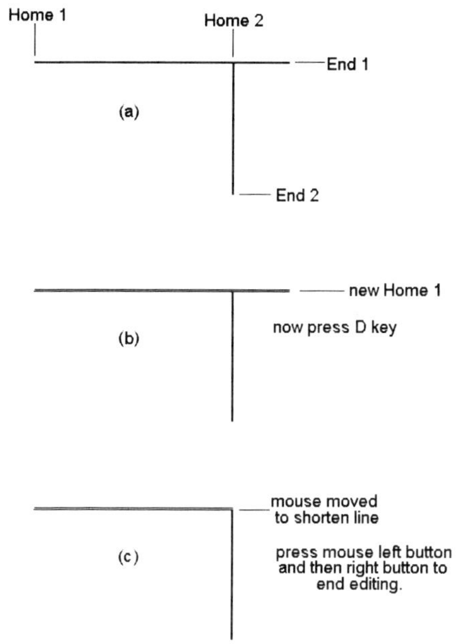

Figure 5.11 Editing on tracks. The stub (a) is a mistake, and editing this line by using the H key (b) allows this to be deleted. The mouse can then be moved to join the lines correctly (c).

to anchor it (c), and then press the ESC key (or right-hand mouse key) to leave the editing action.

- When you use the D key and find that the wrong end of a line moves (with rubber-banding on) then escape and use the Invert key (↑I) before trying the D key again.

Another type of problem is illustrated in Figure 5.12. This is one line with several nodes (a), and we want to see what happens when one segment is deleted. The drawing shows the nodes named, and the action we want to look at is the use of the D key when node 2 is deleted. This is illustrated in (b) – the heavy line of a new track appears, joining node 1 and node 3, with the thin rubber-band lines of the original showing. These thin lines do not show unless rubber-banding is switched on.

If you do not move the mouse, and press RETURN or the left-hand mouse button, the original track pattern will be restored. If you press the ESC key or the right-hand mouse button this will become the fixed track line (c). Until you press the key or button, you can move the cursor as you want to make a new track path between nodes 1 and 3.

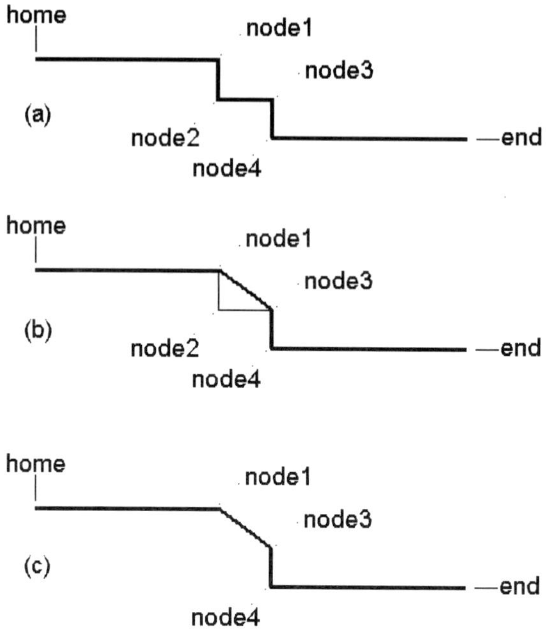

Figure 5.12 The line (a) has several nodes, and deleting node 2 (b) will alter the line shape with node 1 and node 3 joined directly (c).

- Note that node 2 has been removed in this example. When a segment is deleted, the node nearer the home end is removed.

Reshaping a track line is always easier when the line consists of several segments, allowing you a choice of nodes to delete. If you want to reshape a line that consists of only one segment, or with no nodes near where you want to alter the line, it is much simpler to Zap the line (using the ↑Z key) and start again.

You can also insert a new node in an existing line that consists of several segments. Once again, this is easier to use if you first switch on rubber-banding with the X key. Place the cursor on the line near where you want the new node and press the F1 key. You will see the rubber-band line appear from whatever point you move the cursor to, inserting a new node between the existing ones on the line where you have placed the cursor, Figure 5.13. This diagram shows the successive steps of placing the cursor (a), using the edit key (b), moving the cursor as required (c), and then fixing the new line with the mouse button or RETURN key (d).

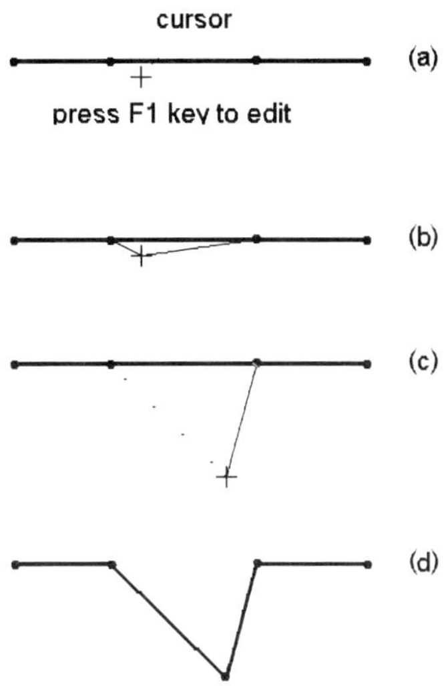

Figure 5.13 Inserting a new node with rubber-banding switched on. The cursor is placed off the line (a) and clicking will show the rubber-banded outline (b). This can be pulled out (c) and anchored to provide the new node position (d).

Using vias

When a track has to pass from one side of a board to the other a via is required. A via is a hole drilled from a pad on one side of the board to a pad on the other, metallized, and sometimes (but not necessarily) filled with solder so that it provides a conducting path between the pads. A via may not be needed if you can use the mounting pads for components to provide conducting paths by way of the component pins.

EASY-PC possesses an auto-via option which is switched off and on in the User Defaults screen (F2) by pressing the P key. When auto-via is switched on, a track that is made in the alternative layer (usually layer 8, in blue) will optionally have a via placed at the point of intersection. In other words, if you join two tracks on layer 0 with a track on layer 8, the vias will be provided for you. This is very useful, but you need to practise it.

Figure 5.14 shows an imaginary section of PCB with several lines that will be crossed by horizontal lines. We do not want these lines to meet, so that the crossings will be on the other side of the board. This has been drawn with the crossing segments about to be put in. Before the crossing lines and the remainder of the horizontal lines are drawn, the User Defaults menu is checked, Figure 5.15, to ensure that option P (Auto-Via) is ON.

Before the first crossing line is drawn, the cursor is placed at the starting point, which is the end of one horizontal line, and the F1 key is pressed. The central menu is used to select Seg. Layer (or you can press the L key). A

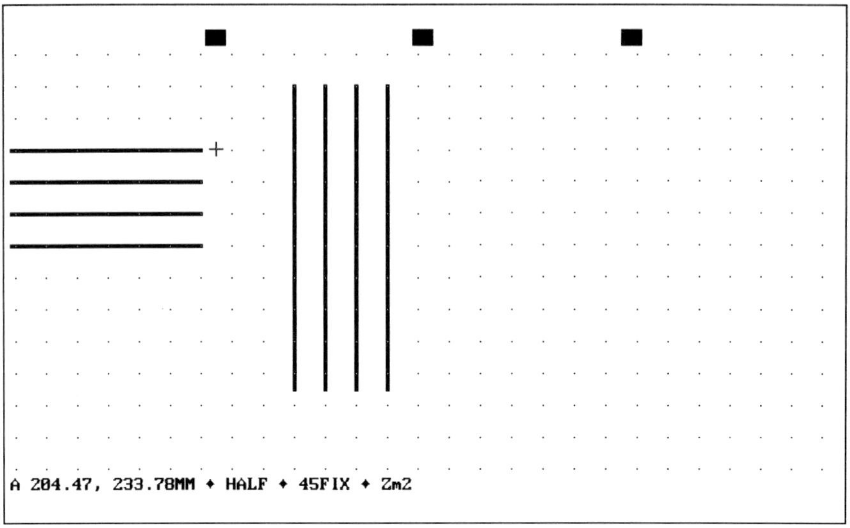

Figure 5.14 A section of board with tracks about to cross. This will be avoided by using the opposite layer.

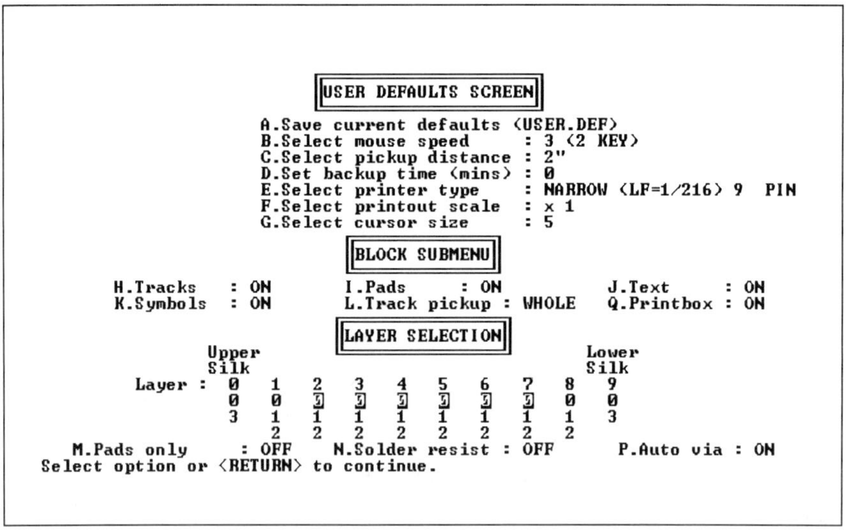

Figure 5.15 Checking the User Defaults screen to ensure that Auto Via is switched on.

message at the foot of the screen asks you to key in the layer, and since you want this track to be on the opposite side from the existing tracks you press the 8 key for layer 8, the opposite side of a double-sided board.

You will then see a message asking if you want to insert a via, and reminding you that, since auto-via is ON, you need only press the RETURN key to insert the via. Doing this (or pressing the left-hand mouse button) will insert the via and make the new track in blue as you move the mouse. You can repeat this at the other side, using layer 1 to meet up with the existing track, Figure 5.16. Figure 5.17 shows the entire set of lines drawn with the layer changes and vias.

- Note that the automatic via insertion operates only when you make a change during drawing a line. You cannot, in the above example,

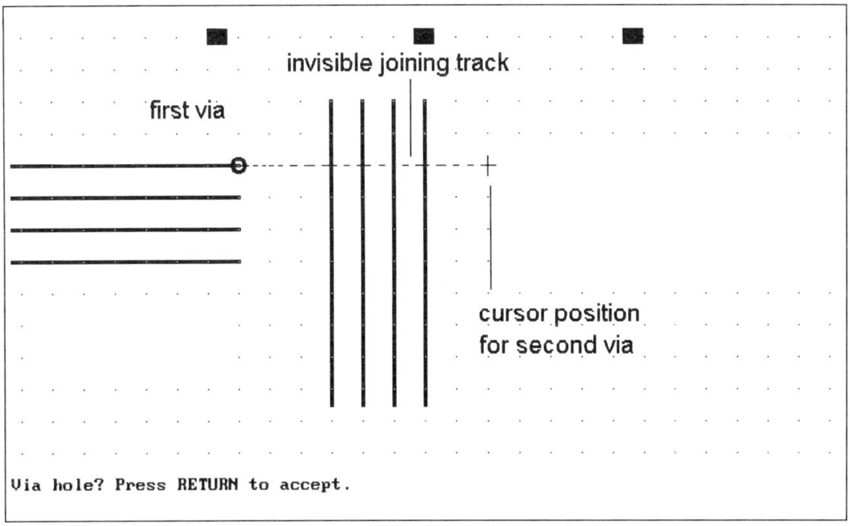

Figure 5.16 Inserting the via automatically. When the layer is changed at a node, you will be asked if you want to insert a via.

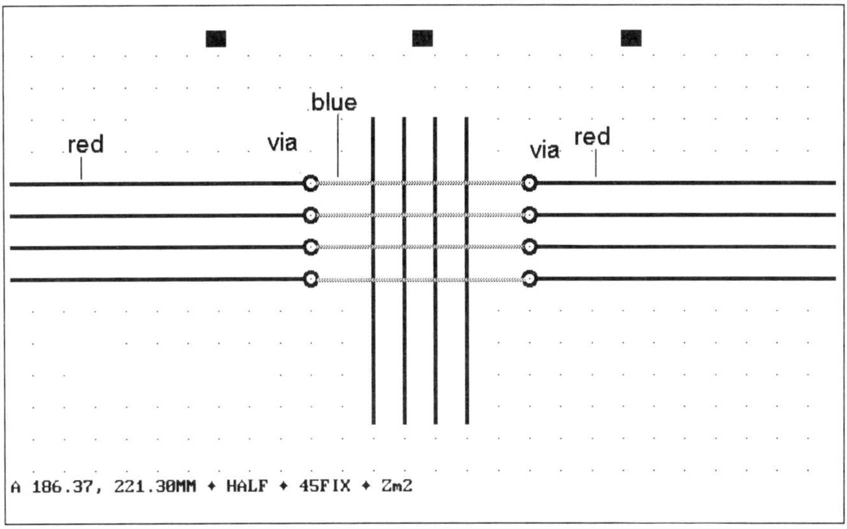

Figure 5.17 The vias and tracks in place.

position the cursor at the end of one horizontal segment, change layer, and expect to make an automatic via hole. You must start or edit a line in layer 1 before you can make a new node and change layer.

- Don't create unnecessary nodes, because each node takes some memory and the use of several nodes on a line makes it tedious to zap that line. You can also conserve memory by using 90° fix rather than 45° fix.

Repeat

The use of Repeat for symbols, tracks, text and pads is a useful way of avoiding a lot of key pressing. The command format is the same for all three cases:

1 Put the cursor on to the symbol, pad, text or track you want to copy.

2 Press the appropriate edit key (F7, F3, F5 or F1).

3 Place the cursor where you want the copy to appear. If you are copying a track you will see the rubber-band outlines if rubber-banding is on. Do not be distracted by these – switch off rubber-banding if you find the appearance a nuisance.

4 Press the R key to make the copy. The home node of the track will be located at the cursor position, not the active node.

Remember that you can also use the S key if you want to shift an object (symbol, pad, text or track).

Relative co-ordinates

Sometimes when you copy or shift an object, particularly a pad, you need to measure the precise distance from another point, perhaps another pad. The co-ordinates that you normally see at the foot of the screen, at the left-hand side are, by default, absolute co-ordinates which take as their starting point the bottom left-hand corner of the 17" x 17" board that is also a default.

You can alter this if you need to, selecting any point as an origin and making the co-ordinates relative.

1 Place the cursor where you want the new origin to be.

2 Press the O key to make this the new origin.

3 Press the ↑R key to make co-ordinates read relative to the new origin.

  - Note that when you do this, you can expect to see negative co-ordinates appear when the cursor is placed to the left or below the new relative origin.

6 Illustrated summary

This summary of EASY-PC actions has been put, for convenience, in alphabetical order. It assumes that you have read the preceding chapters, so that instructions concerning the placing of the cursor and the use of the editing keys do not need to be repeated in detail. In these entries, the **Use:** item refers to when the instruction can be used, since some instructions can be used only after one of the F-keys has been pressed. An 'Any time' entry means that the action can be obtained at any stage in a layout or drawing. **Key:** means the key that can be pressed (assuming that you are using the appropriate menu if this is not an 'any time' item) to obtain the action as an alternative to using the drop-down menus. Items are illustrated as required – if the previous chapters contain illustrations of the action then this is repeated only if it saves having to hunt through the pages for the information. The ↑ symbol is used to indicate the SHIFT key.

Attributes, Current Item

Use: While using any editing mode.

Key: Grey + (on number pad).

Action: The attributes (size, colour, layer etc.) are displayed at the bottom of the screen while the + key is held down.

Angle, Block

Use: When a block has been drawn.

Key: C or A.

Action: The whole block can be rotated clockwise (C) or anticlockwise (A) in a step of 90° each time the key is pressed.

Angle, Symbol

Use: When editing a symbol.

Key: A.

Action: The symbol can be rotated in 90° steps. After pressing the A key you can enter the numbers 1 (90° anticlockwise), 2 (180° anticlockwise) or 3 (270° anticlockwise). The 0 key will return the symbol to its original angle.

Angle, Text

Use: When editing text.

Key: A.

Action: The text object can be rotated in 90° steps. When the A key is used you can enter the numbers 1 (90° anticlockwise), 2 (180° anticlockwise) or 3 (270° anticlockwise). The 0 key will return the text to its original angle.

Angle, Track

Use: Any time.

Key: ↑A.

Action: The angle between any two drawn tracks can be set as 45FIX (45° or 90° angles), 90FIX (90° only) or MULTI (any angle). These options will appear in turn as the ↑A key is used. Avoid MULTI if at all possible.

- When a segment is deleted, the automatic joining of nodes can cause lines to appear which are not at 45° or 90° angles. These should be edited to make tracks at the normal angles.

Arc (Circle)

Use: After a circle has been drawn, with cursor on circle.

Key: Track edit F1.

Action: Use the H key to find the starting point of the circle, and the N key to get to the start of the portion you want to remove. Then use the D key to delete segments in turn (moving first anticlockwise and then clockwise) until you are left with the arc you want. Figure 6.1 shows the steps as they appear on screen – the small lines at the nodes will disappear when you use the ESC key or right-hand mouse button.

- A circle consists of a set of small segments, which is what makes this editing method possible. You should practise the use of this action on a separate drawing before you attempt to make serious use of it. See also the notes in Chapter 3 regarding the use of resolution options. Use Auto unless you have a pressing necessity to use High or Low resolution.

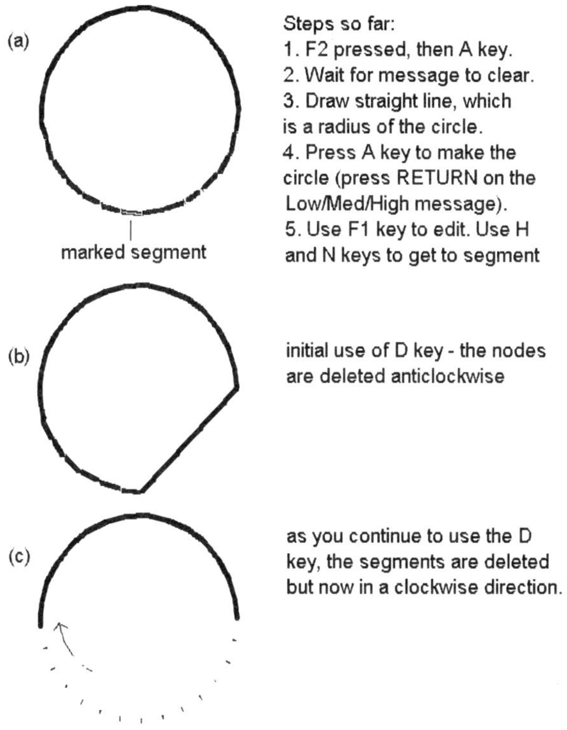

Figure 6.1 Editing a circle into an arc using the Track menu.

Autosave

Use: When User Defaults screen is visible (↑F2).

Key: D.

Action: Sets the time interval between autosaving of the current drawing. As the D key is pressed, the autosave interval alters in increments of 5 minutes from 0 to 30 minutes. A useful rule of thumb for the optimum Autosave interval is to make it ten times the amount of time needed to save your file normally. For example, if it takes one minute to save a file, the Autosave interval should be set to 10 minutes.

- This is a valuable safeguard against the loss of a file which might represent a considerable amount of work. You are prompted by the message: `Save new version (Y/N)` when an autosave is due. You should not disable the Autosave function unless you are practising with trivial examples. Remember that Autosave does not make back-ups on a floppy disk, something that you should also do at regular intervals.

Block

Use: Any stage of drawing or editing.

Key: F10 with cursor placed on corner of block.

Action: Allows you to draw a rectangular outline by dragging the mouse. The corner point is placed at the cursor position when you press the F10 key, and when you have dragged the other corner out to where it is needed, pressing the mouse button (or RETURN key) will fix the block outline. Everything within the block outline will then be affected by block commands, as modified by your settings of User Defaults. Note that a block is essential if you intend to print using a dot-matrix printer.

- You should always save a file before using a block command.
- The origin of a block is not a fixed position – it is the position where you start drawing the block and is therefore at one corner.
- You can erase everything inside a block by using ↑Z, the Zap command. You will be prompted to confirm (Y/N) before the action is carried out. Track Pickup in User Defaults (↑F2) must first be set to WHOLE.

Border

Use: Any time.

Key: ↑B.

Action: Switches between showing the outline of the drawing area (the default) or not. The Border should be left On. (The border outline is usually visible only at the higher zoom numbers.)

Circle

Use: When creating track (F2).

Key: A.

Action: The starting cursor position is used as the centre of the circle. Press the F2 key and then draw a piece of track whose ending node will be on the rim of the circle. Press the A key to create the circle. You will be asked about resolution – (L)ow, (M)edium, (H)igh or Auto, with the Auto default being used if you simply press the RETURN key or the left-hand mouse button.

Use the Auto resolution option. A circle is drawn as a large number of straight segments and Auto will ensure that the circle is drawn with as many segments as will make it look circular. If you insist on using High resolution for a large number of circles you will probably run out of memory, and the screen refresh action will take a noticeable time.

- When you draw the line from centre to edge to create your circle, make sure that this is one straight segment, otherwise the A key will have no effect. If you want to draw this line at an odd angle (for example, to draw an arc at an angle) first select Multi-angle, or use Back and Delete to remove the extra node.

Clear All

Use: With unwanted drawing on screen.

Key: ↑F3.

Action: Clears the current drawing on screen so you can start a new drawing of the same type. This is faster than using ↑F10 to return to the main menu.

- You will be prompted to confirm this by the message: `Initialize artwork? Confirm (Y/N)` at the foot of the screen.
- Clear All can be used when you have loaded in a dummy drawing. By editing objects on the dummy drawing you can set the attributes for the new drawing you want to create, and you can start to create this drawing after clearing out the dummy.

Colours

Use: Any time.

Key: ↑F.

Action: Cycles through five sets of foreground colours for the main layers. The background colour is always black. The default set of colours is Red, Blue and White, used for track layer 1, track layer 8 and silk-screen respectively. The other options are typically Red, Green and Brown; Light Red, Light Green and Yellow; Magenta, Blue and Grey; Light Magenta, Light Blue and White. The colours cycle to the next in turn on each press of the ↑F key. The actual colours that you will see depend on whether you are using a CGA or EGA screen mode (these modes are available on VGA screens).

Default Pads

Use: Any time.

Key: ↑F1

Action: The key brings up the pad and track default menu. The default pad width is 0.010", width 1, and the default set of pad widths ranges up to 0.202", width 15. The width 16 pad is used for the automatic via connections. The other pad default is that pads are placed on all layers. The default pad width table is illustrated in Figure 6.2.

- You can create your own pad width defaults and set these as defaults for future use. The current pad is marked by cursors which can be moved using the cursor keys. At each pad number, you can alter the size by using the + or – keys. These values can be fixed for future use by selecting Save Current Defaults in the User Defaults menu (↑F2 key). You will not be allowed to change to ridiculous values (inner width greater than or equal to outer width).

Pad widths in 0.001" units

	Pad No.	Outside	Inside
	1.	10	6
	2.	15	6
	3.	19	15
	4.	23	15
	5.	27	15
	6.	31	15
	7.	35	27
	8.	44	31
	9.	48	31
	10.	60	31
	11.	69	31
	12.	77	31
	13.	85	35
	14.	152	69
	15.	202	123
Auto via :	16.	52	31

Figure 6.2 The Pad Width menu with dimensions in thousandths of an inch.

Defaults, Track

Use: Any time.

Key: ↑F1.

Action: The key brings up the pad and track default menu. The current track width is marked by cursors which can be moved using the cursor keys. At each track number, you can alter the size by using the + or – keys. These values can be fixed for future use by selecting Save Current Defaults in the User Defaults menu (↑F2 key).

Defaults, User

Use: Any time.

Key: ↑F2.

Action: Allows you to set user defaults as illustrated in Figure 6.3. Each is selected using a letter key, and most of the items switch value each time the key is used. These values can be fixed for future use by selecting Save Current Defaults. The values of these defaults are saved with a drawing file.

- Typing a number key will alter the state of the layers 0 to 9. You can hide or display each layer, or change its colour. You can also control the output to a dot-matrix printer from this menu.

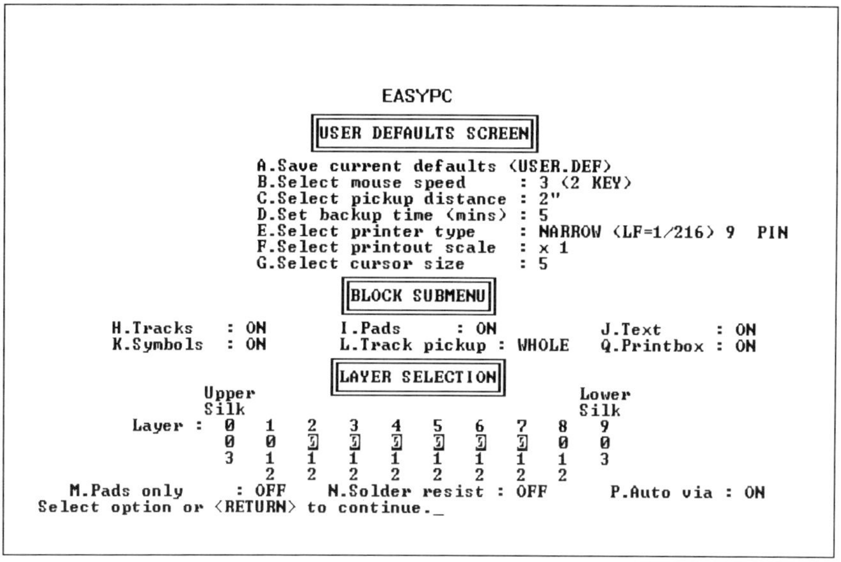

Figure 6.3 The User Defaults screen.

Delete Block

Use: When a block has been drawn.

Key: ↑Z.

Action: will delete all the text, symbols, tracks and pads enclosed in the block. You will be asked to confirm (using Y or N key) because there is no undo action – once a block has been deleted it will stay deleted unless you have the old version on file. If the block contains tracks, Track Pickup in the User Defaults screen (↑F2) must first be set to WHOLE.

Delete Pad

Use: When pad edit (F3) key has been used to enter pad edit.

Key: D.

Action: Will delete the pad that is currently being edited.

Delete Symbol

Use: When editing symbols (F7 action).

Key: ↑Z.

Action: Will delete the symbol that the cursor was on when the F7 key was used followed by ↑Z. You will be asked to confirm with Y or N key.

Delete Text

Use: When editing text (F5 action).

Key: ↑Z.

Action: will delete the current text that is being edited. You will be asked to confirm with the Y or N key.

Delete Track

Use: When editing track (F1 action).

Key: ↑Z.

Action: Will delete all of the current track from home node to end node. You will be asked to confirm with the Y or N key. Note the difference between this and the segment (node) delete action.

Drawing Defaults

Use: Any time.

Key: ↑F1.

Action: Brings up the Track and Pad Widths screen. You can alter all of the default track widths and pad widths, using the cursor to select and the + and – keys to alter values. The values cycle around a set of 128, so that all values can be used – you cannot set an impossible track width nor make the outer width of a pad greater than its inner width figure.

- These defaults are stored along with any layout that uses them, and will appear when that layout is loaded and edited again. You can also make these settings the defaults for any future layouts by using the Save Current Defaults option in the User Defaults menu (↑F2 key).

Edit Pad

Use: With cursor on a pad.

Key: F3.

Action: Allows you to move, copy or delete a pad, with options to alter layer, width and shape (variant). The move action is done by moving the cursor and clicking the left-hand mouse button (or pressing the RETURN key). Other choices are obtained from the central drop-down menu or from letter keys and must be used before the pad is fixed in place with the right-hand mouse button or ESC key.

Edit Symbol

Use: When cursor is over a symbol.

Key: F7.

Action: Allows the symbol to be moved by moving the cursor and clicking with the left-hand mouse button (or pressing the RETURN key). Other actions are carried out using central drop-down menu or keys. The other actions are Edit Text (T), Current (C), Layer (L), Angle (A), Flip (F), Repeat (R) and Zap (↑Z).

Edit Text

Use: When cursor is over text.

Key: F5.

Action: Allows the text to be moved by moving the cursor and clicking with the left-hand mouse button (or pressing the RETURN key). Other actions are carried out using central drop-down menu or keys. The other actions are Edit Text (T), Current (C), Layer (L), Size (S), Width (W), Angle (A), Flip (F), Repeat (R) and Zap (↑Z).

- Use the Edit Text (T) item to alter the context (text characters) rather than altering the text position.

Note that you can put in the usual non-ASCII characters by using the ALT key and typing a number between 128 and 255. For example, you can get the é character by holding down the ALT key and typing 130 on the number keypad, then releasing the ALT key.

Edit Track

Use: When cursor is over a node of a track.

Key: F1.

Action: Allows the track to be re-routed by making new nodes – with rubber-banding on, move the cursor and click the left-hand mouse button (or RETURN key) to fix the new route. The new route may not use the normal 45° and 90° angles, so this should be done with care. The other actions, obtained from the central drop-down menu or keys are Home (H), Back (B), Next (N), End (E), Arc/Circle (A), Current (C), Delete (D), All-Layer (↑L), Seg. Layer (L), All Width (W), Seg. Width (↑W), Repeat (R), Invert (↑I), Flip (F), Shift (S) and Zap (↑Z).

- The Invert (↑I) command, which reverses the position of home and end nodes, can be used several times in succession, with rubber-banding switched on, to find the best way to proceed with an edit. The Flip (F) command selects each of two possible routes with fixed angles.

Flip Block

Use: When a block has been defined.

Key: F.

Action: Mirrors a block around a vertical line passing through the cursor (see Figure 6.4). One use of this is to create mirrored schematic components, as for flip-flop diagrams or symmetrical PCB layouts. Another is to allow

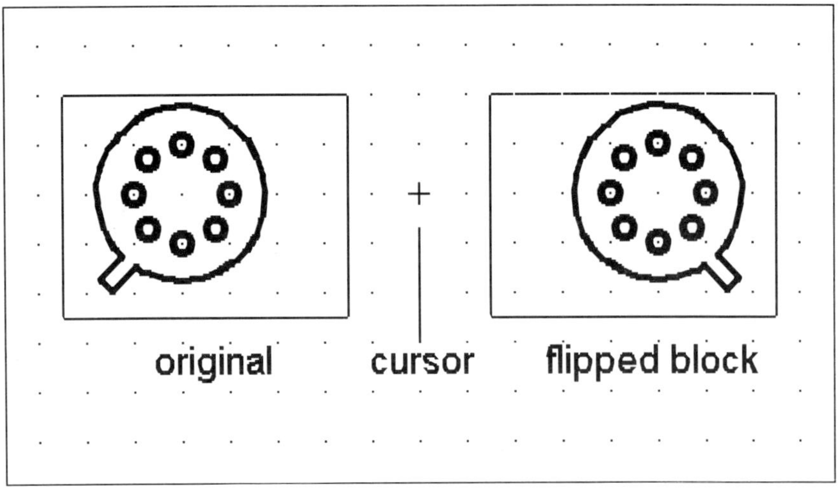

Figure 6.4 Using the Block Flip command.

you to look on the bottom layer of a board as if it were on top, so that you can print the copper layer for this side.

- You should not use a block flip on an IC PCB position, as it will alter the relative pin positions – use a rotation if you want to turn ICs around.

Flip Text

Use: While editing or entering text.

Key: F.

Action: Allows text to be placed on the back layer in mirrored form so that it will read correctly when the silk-screen for the back is printed. You can restore flipped text by using Text edit and repeating the Flip command.

- Note that you can also enter text on a copper layer for identification purposes. Such text should be placed clear of tracks, pads and symbols.

Grid

Use: Any time.

Key: G.

Action: Switches the grid dot display on or off. You should work with the grid on until a drawing is complete, so that you can be sure that objects are correctly lined up. You can turn the grid off for a last check on a layout or schematic unobscured by the grid dots.

- The grid positioning also determines the snap positions, but the snap action will take place even if the grid is not visible. The Scale command (↑G) will show a ruler display in place of the grid.

Large Cursor

Use: Any time.

Key: ↑C.

Action: Switches the cursor between its small cross and large cross shapes.

- The large cross is very useful if you have to line up objects that are some distance apart.

Layer

Use: With track, symbol, text or pad.

Key: L.

Action: Determines which layer will be used for the next action. The layers that are normally used are 0 for silk-screen printing of text on the upper surface of a PCB, 1 for top surface symbols, tracks and pads. Layer 8 is used for underside symbols, text, tracks and pads, and 9 for underside silk-screen text. Copper text on the underside is often used to locate test points.

- When a track is switched to a different layer (usually 1 to 8 or 8 to 1) you can use the option of inserting a via automatically.
- Layers 2 to 7 are intended for multi-layer boards, and you can also use them with ordinary boards to carry information such as drilling instructions.

Load

Use: In Filing System menu (F9 key).

Key: L.

Action: Loads a file whose name must be typed followed by RETURN key. The extension letters of PCB or SCH do not need to be typed because these are automatically inserted according to whether you are working with a layout or a schematic.

- If you want to see a directory of names, use the D key. You can also use ↑P to set a path for locating files.

Load Pad

Use: During PCB layout or schematic.

Key: F4.

Action: Places a pad of the current width and other attributes at the cursor position. If you want to alter width or shape of pad, you must do so before pressing the right-hand mouse button or the ESC key. You can move the pad using the RETURN key or left-hand mouse button, before you press the ESC key (or right-hand mouse button).

Load Symbol

Use: During PCB layout or schematic.

Key: F8.

Action: Requests symbol name and provides default library name of SYMB.LLB for PCB layout or SYMB.SLB for schematic. When the RETURN key or left-hand mouse key is pressed this will place the symbol with its part-origin at the cursor. The symbol can be moved to a new position by moving the cursor and pressing the left-hand mouse button (or RETURN key) and this can be done until the ESC key or right-hand mouse button is pressed.

- If you need several copies of one symbol, load one, move the cursor and use the R key to place the others.

Merge

Use: While working on layout or schematic.

Key: ↑M.

Action: Will load in another file of the same type (PCB or SCH), but without erasing the existing layout or schematic. A normal load will always clear the drawing space – you will be prompted to save any drawing that has not been saved since being changed.

Merge will place the merging file in the same position on the drawing area as it was saved from. You should therefore save such files from a standardized position (such as top right-hand corner) and use the opposite corner for your drawing work. This allows you to merge in another drawing without overlapping the work you are currently engaged on. Always save your original file before using a Merge in case you have made a mistake in positioning the drawings.

Move Along Track

Use: When editing track (F1).

Key: H (Home), E (End), N (Next) and B (Back); also C (Current Node).

Action: Allows you to see the current segment and node marked for editing actions. The current node is the end of the current segment which is closer to the home node. Using the C key will move the cursor to the current node, no matter where the cursor had been placed.

- You can use the ↑I facility to interchange home and end nodes so as to work from the other end of the track.

New Text

Use: On schematics or on the PCB silk-screen layer.

Key: F6.

Action: Allows you to place text on the silk-screen or any other layer. The default layer is 0, and text placed on the other layer (9) should be flipped so that it will read correctly when this layer is printed. The text can be moved or edited before the ESC key or right-hand mouse button is used to anchor the text. When you edit text, the attributes of that text will be used for the next new text that you enter. This allows you to keep text samples in a dummy drawing and to pick them up by editing so as to use on a new drawing.

- Identifying text can also be placed on a copper layer.

Node Delete

Use: In track editing mode after H, N, B, E keys have been used to locate a segment.

Key: D.

Action: Will delete the node nearer the home end of the track, and will create a new segment joining the adjacent pair of nodes, one on either side. See Chapter 5 for illustrations of this action. When a node is deleted, the information on items such as track width and layer that were stored at the node position will be lost, and the information will be provided by the other nodes.

- You should carry out this deletion action with rubber-banding switched on so that you can see the line of the new track that will form the replacement segment. If you do not use rubber-banding you may find that the replacement segment is not where you expect it. With rubber-banding on it is easy to see where you need to place a replacement node, or where to place the cursor to edit a line.
- You cannot break a line except by deleting from one end.

Node Insert

Use: During track editing (F1 key).

Key: RETURN or L-H mouse.

Action: With the current node highlighted, place the cursor outside the segment and click to make a new node and join to the nodes of the current segment.

- This is much easier to use if rubber-banding is switched on, key X.

Origin, Block

Use: With block selected.

Keys: O to locate origin at cursor, ↑R to alter origin.

Action: The origin (co-ordinate position 0,0) for a block is the corner at which the drawing of the block started. You can switch between relative and absolute co-ordinates by using the ↑R key.

Origin, Drawing

See *Origin, Block* above. Note that block operations use this origin so that it will have to be repositioned afterwards.

Pad, New

Use: When working on layout or schematic.

Key: F4.

Action: Places a default width-1 pad at the cursor position. You can move this pad by moving the cursor and clicking the left-hand mouse button or pressing the RETURN key. If the shape or width is to be changed, this must be done before the right-hand mouse button or ESC key is used. Once a pad is placed or edited, its attributes will be used for the next pad that is put in.

Pan

Use: At any time.

Key: P.

Action: Will shift the current display so that its centre is around the point of the cursor when the P key was pressed.

- The Pan action also redraws the screen, so that this action, with the cursor placed at the centre of the drawing, is useful for removing any clutter left by editing actions.

Partial Track Pickup

Use: In User Defaults menu (↑F2).

Key: L.

Action: Allows the block to be moved or rotated while maintaining the connections to tracks outside the block (see Figure 6.5). Also prevents zapping (↑Z) of blocks that contain lines or tracks.

- These track connections are likely to take up angles other than the normal 45° or 90°, and track editing is likely to be needed.

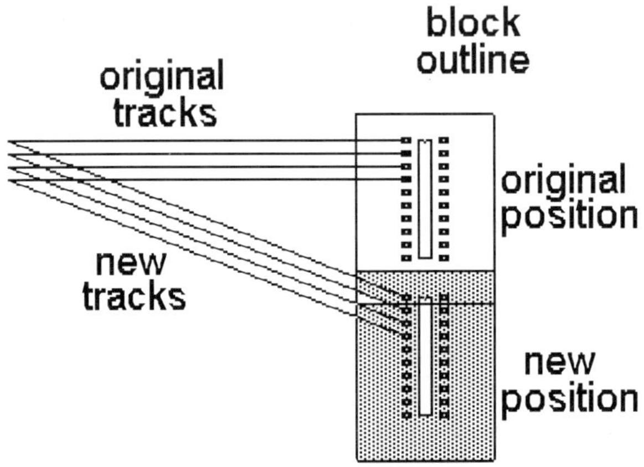

Figure 6.5 Moving a block with Partial Track Pickup.

Repeat Block

Use: With block outlined.

Key: R.

Action: Will make a copy of the block which will appear at the cursor position. This is very useful for copying symbols, text, pads and track layouts. Track Pickup must be set to Whole in the User Defaults menu.

Repeat Pad

Use: When pad editing (key F3) or inserting (key F4).

Key: R.

Action: makes another copy of the pad appear at the cursor position. This is faster than loading another pad using the F4 (new pad) command.

Repeat Symbol

Use: When editing symbols (key F7) or inserting symbols (key F8).

Key: R.

Action: Makes another copy of the symbol at the current cursor position.

Repeat Text

Use: When text editing (key F5) or inserting (key F6).

Key: R.

Action: Makes another copy of the text at the cursor position. Not so widely used as the other repeat actions because you seldom want to copy text identically. You might, however, have a set of text items such as TRSFIL1, TRSFIL2, TRSFIL3, in which case it is quicker to type one and use Repeat Text, then edit the numbers.

- When you want to copy a piece of existing text by using repeat, make sure that you pick up the text for editing by placing the cursor at the origin, which is at the start of the text. If you copy a block that contains text, make sure that the origin of the text is included in the block.

Repeat Track

Use: When track editing (key F1) or insertion (key F2).

Key: R.

Action: Makes a copy of the track at the cursor position. The home node of the track will be placed at the cursor position. This action is ideal for creating bus lines in microprocessor circuits.

Rotate Block

Use: With a block selected.

Key: C or A.

Action: The C key will rotate the block trough 90° clockwise; the A key will rotate through 90° anticlockwise. The rotation is always around the cursor position.

- You should use Partial Track Pickup in User Defaults in order to maintain connectivity.

Rotate Symbol

Use: With symbol being edited (F7 key).

Key: A.

Action: You must enter a number 0 through 3 to define the number of 90° steps of rotation, which is always anticlockwise.

- You can rotate, flip and rotate again to achieve the effect of flipping around a horizontal axis.

Rubber-banding

Use: When laying track or schematic lines.

Key: X.

Action: Shows white lines following the cursor position to indicate the effect of moving the cursor. These lines change to the default track or line colour when the mouse is clicked or RETURN pressed.

- Rubber-banding should preferably be switched on for all track laying and editing actions, because it makes the end result of a track action much clearer.

Save

Use: From Filing menu (F9 key).

Key: S.

Action: You will be asked for a filename of no more than eight letters. This must not conflict with any name currently in use for a file of the same type (PCB or SCH). When a file has been saved several times, an older version of the file will also exist on the disk. For example, if you have saved files called MEMBANK.PCB and MEMBANK.SCH several times you will find files called MEMBANK.BK1 and MEMBANK.BK2 also in the same directory. The file with the extension BK1 is the MEMBANK.PCB backup, and the file with the extension of BK2 is the MEMBANK.SCH backup. If you find that your most recently saved version is faulty you can rename the appropriate backup file, using a new name and the PCB or SCH extension, to use in EASY-PC.

- If you use the Exit (↑F10) or Clear All (↑F3) command on work that has not been saved since it was created or modified, you will be reminded and asked if you want to save the file. You will then be asked again if you want to abandon it.

Scale

Use: Any time.

Key: ↑G.

Action: Places ruler markings around the screen as an aid to placing objects. This is an alternative to using the grid, and the two cannot appear together. The Scale is best used along with the large cursor (↑C key) as illustrated in Figure 6.6. The scale markings are at fixed 0.1" intervals.

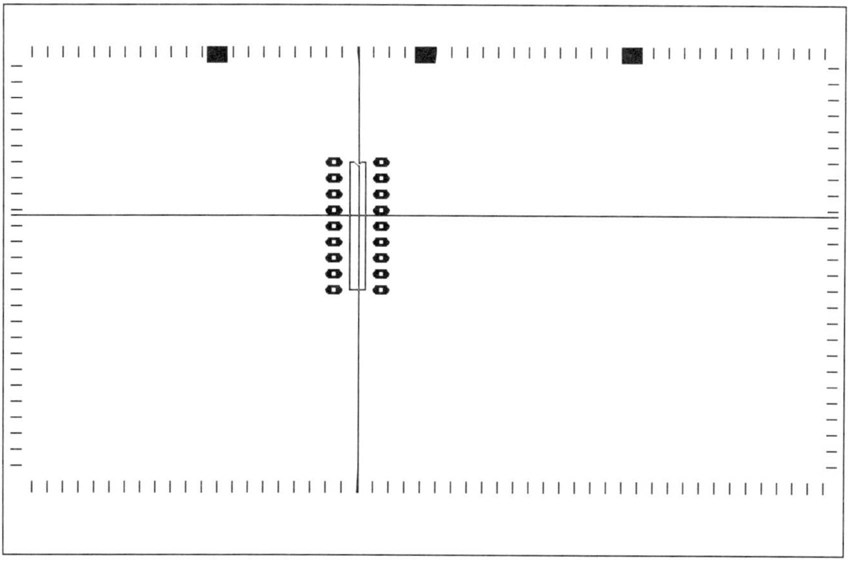

Figure 6.6 Using the Scale along with the large cursor.

Shape of Pad

Use: From Edit Pad (F3) or New Pad (F4).

Key: V.

Action: The V menu shows the set of pad shapes, Figure 6.7, from which you can select by typing the letters as shown. When you put the cursor on the central square menu, you will see the Variant menu appear when you move the cursor to the Variant line – the shapes and their associated letters appear in a column.

- The selection of pad shape (and width) must be made before the pad is fixed in place by using the ESC key or right-hand mouse button.

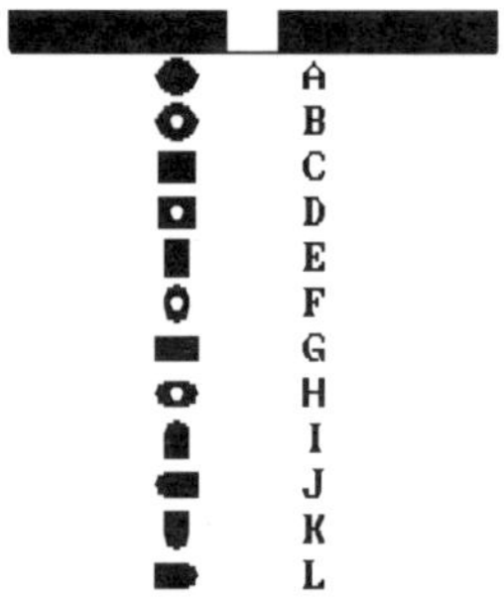

Figure 6.7 The selection of pad shapes obtained from the Variant menu (V key)

Shift Block

Use: When a block outline has been created and fixed.

Key: S.

Action: The use of the S key will move the origin of the block and its contents to the cursor position – you must position the cursor before using the S key. Remember that the origin of the block is the corner at which you started drawing the block.

- If the block crosses tracks you should decide whether to use *Whole* or *Partial* track pickup.

Size, Text

Use: When editing or entering text (F5 or F6).

Key: S.

Action: Will alter the height of new or existing text with a choice of four sizes. The size is selected by typing a number:

Number	Text Height
1	0.05"
2	0.1"
3	0.2"
4	0.4"

Size of Pad

Use: When entering or editing pad (F3 or F4).

Key: W.

Action: Will allow you to change the pad width by using a typed number, one of the set in the User Defaults table. (See the table in Figure 6.2.) If you have changed the sizes in this table, the new sizes will be used. See Appendix E for suggested defaults for different types of PCB work.

- The pad width cannot be changed once the ESC key (right-hand mouse button) has been used. If you want to make changes after using ESC, place the cursor on the pad and press F3 to edit that pad again.
- Remember that you can see the pad attributes if you start an edit and then press the grey + key.

Snap

Use: When placing or editing all objects.

Key: M.

Action: Ensures that objects are precisely placed relative to the grid. Snap points can be located on the grid points (GRID), at half-grid intervals (HALF), at quarter-grid intervals (QTR), Freehand (FREE) or Near (NEAR). Pressing the M key will cycle through these options. The current snap mode is always displayed at the foot of the screen while you are drawing or editing.

- In the near-point mode the snap will be to the nearest node on a track or to the origin on a symbol or pad. You cannot start a drawing in Near mode – you will get a message to the effect that there is no point within range.

Status

Use: Any time.

Key: ↑F4.

Action: Provides a screen print of the current state of a drawing, showing the number of tracks, symbols, text objects and pads, also percentage of memory used and reminders of file and library. Figure 6.8 shows an example.

Text Angle

Use: When editing text.

Key: A.

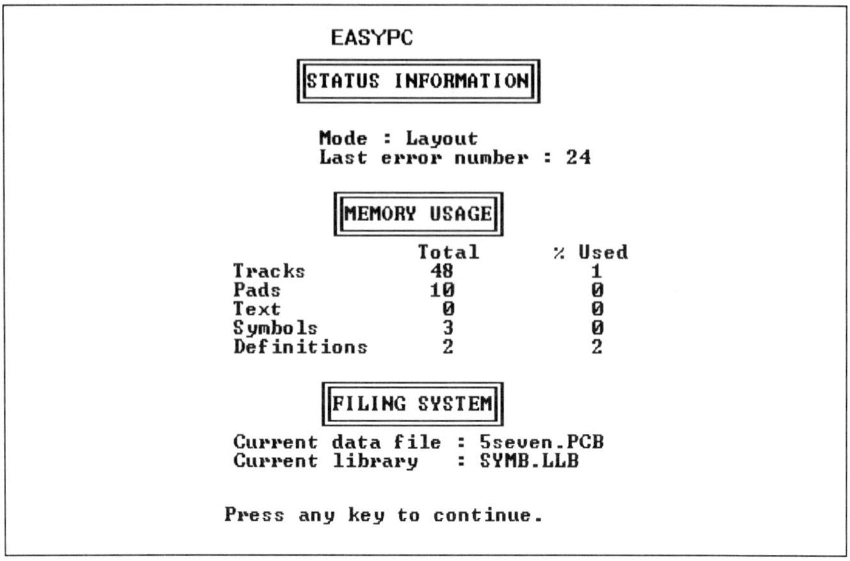

EASYPC

STATUS INFORMATION

Mode : Layout
Last error number : 24

MEMORY USAGE

	Total	% Used
Tracks	48	1
Pads	10	0
Text	0	0
Symbols	3	0
Definitions	2	2

FILING SYSTEM

Current data file : 5seven.PCB
Current library : SYMB.LLB

Press any key to continue.

Figure 6.8 The Status Information form for a layout.

Action: The text object can be rotated in 90° steps. When the A key is used you can enter the numbers 1 (90° anticlockwise), 2 (180° anticlockwise) or 3 (270° anticlockwise). The 0 key will return the text to its original angle.

Track Angle

Use: Any time.

Key: ↑A.

Action: The angle between any two drawn tracks can be set as 45FIX (45° or 90° angles), 90FIX (90° only) or MULTI (any angle). These options will appear in turn as the ↑A key is used. Avoid MULTI if at all possible.

- When a segment is deleted, the automatic joining of nodes can cause lines to appear which are not at 45° or 90° angles. These should be edited to make tracks at the normal angles. Use 90FIX for schematics.

Units

Use: Any time.

Key: I.

Action: Switches between imperial and metric units for cursor co-ordinates.

- Measurements such as pad and track widths continue to be quoted in imperial units whatever units are being used for co-ordinates.

Unzoom

Use: Any time.

Key: U.

Action: Switches to the next lower magnification (corresponding to a higher zoom level).

User Defaults

Use: Any time.

Key: ↑F2.

Action: Displays the User Defaults menu, see Figure 6.3. Changes made in this menu will affect what you do from then on, and if you want to fix these values for future use, then select Save Current Defaults before you leave the User Defaults menu.

- If defaults are changed only for the duration of one layout or drawing they will be saved with that file and will be available when the file is being edited.

Vias

Use: In track entry, only at a node, F2 key.

Key: None or F4.

Action: If the layer number is changed at a node (key L, then type number) and Auto Via is switched on in User Defaults, you will be asked if you want an auto via to link the tracks on the different layers. Press RETURN or left-hand mouse button to accept, ESC or right-hand mouse button to decline. If Auto Via is off, you can insert a via manually by using pad width number 16 (see Pad Width). The size of the via can be changed (↑F1 key) by editing the size of the number 16 pad. Note that this number provides the smallest via size that can be reliably manufactured – you can alter the User Defaults table for larger vias but you should not attempt to create smaller vias.

- If you need to insert a via in the middle of a piece of track, put a node in at that point.

Whole Track Pickup

Use: In block move or rotate actions.

Key: None.

Action: Track pickup is set in the User Defaults menu (↑F2 key). When Whole Track Pickup is selected, moving a block that contains tracks extending outside the block will not move these tracks. It will only move the other symbols, severing them from the tracks outside the block, see Figure 6.9.

- If you want to maintain track connections, use Partial Track Pickup.

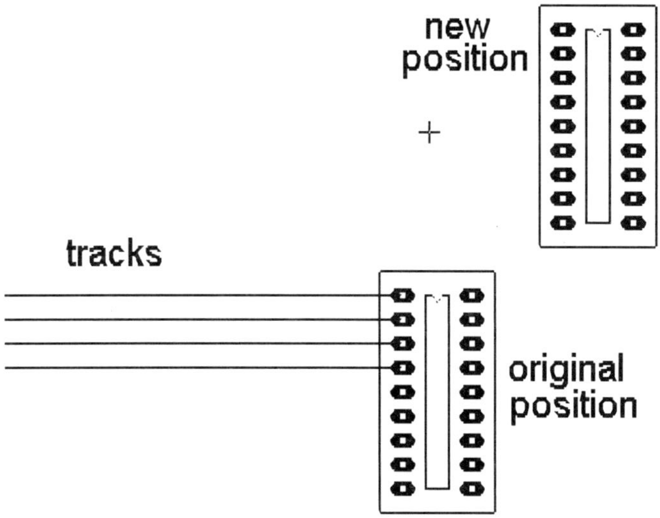

Figure 6.9 Using Whole Track Pickup with a block move – none of the track within the block has been moved.

Width, Pad

Use: When laying down a new pad (F4) or editing a pad (F3).

Key: W.

Action: Will allow you to specify the number for a different pad width, using the figures in the User Defaults table. See Appendix E for typical default values to use for different types of PCB work.

Width, Track

Use: When using new track (F2) or track edit (F1).

Key: W.

Action: Will allow you to alter track width for new or edited track. With the cursor on the track, press the W key and enter the width number you want to use (range 1 to 8). The width is fixed when you use the ESC key or the right-hand mouse button. A table of widths and width numbers is illustrated in Figure 6.10. Note that you can specify width of whole track by using the ↑W key, or the width of the current segment by using the ↑W key. Altering the width of a segment is a way of necking down a track width so that it can be passed between IC pins.

- The largest widths are suitable for ground planes and will cover a large part of a board. These widths should not be used unless you really need them.

Track widths in 0.001" units

number	width
1.	2
2.	6
3.	10
4.	15
5.	52
6.	102
7.	202
8.	460

Figure 6.10 Track widths and their reference numbers. Widths are quoted in thousandths of an inch even if you are working in metric dimensions.

Width, Text

Use: During text entry (F6) or editing (F5).

Key: W.

Action: Alters the width of the lines that are used to draw text, using a number in the range 1 to 8.

Width No.	Width of text
1	0.005
2	0.01
3	0.02
4	0.03
5	0.038
6	0.042
7	0.055
8	0.062

Note that this has no effect on the width of the letters themselves, which remains at whatever setting was last used. The table shows the widths as measured against the screen co-ordinate readings. Wide text should be used along with larger height numbers.

- The minimum width figure of 0.005" is too small for reliable silk-screen printing.

Width, Track

See *Track Width* above.

Zoom

Use: Any time.

Key: Z.

Action: Will zoom to the next larger magnification (smaller zoom number). The largest magnification is zoom 1, the smallest is zoom 7. See also *Unzoom*.

7 Using EASY-PC Professional

EASY-PC Professional is a considerable extension of EASY-PC, intended, as the name suggests, for the professional who is concerned with the manufacture of PC boards on a day-to-day basis rather than the intermittent production of a PCB for making prototypes. The main distinguishing feature of EASY-PC Professional is its ability to convert a schematic into a rat's-nest PCB layout, eliminating the tedious process of making all the track connections manually. It offers considerably more than this, however, with an extended set of commands, new facilities, a different method of treating symbols, faster printing from a laser printer, better-looking schematics, and a host of other features.

The statistics alone are impressive, with a drawing space of 32" x 32" maximum and a resolution of 0.001". You can use up to sixteen layers and each layer can be displayed in a different colour, with colours selectable. The outstanding feature is that once a schematic has been drawn and checked, the operation of the circuit, linear or digital, can be analysed and a PCB with correct connections can be drawn automatically. The link to the linear and digital simulator software is a unique feature of the Professional version.

Because EASY-PC Professional is a large package, but also because it follows closely the methods used by the simpler package, the description in this chapter is condensed and is aimed mainly at illustrating the extra facilities that the Professional version offers.

- If you use Windows on a 386 (or higher) machine, note that EASY-PC Professional, like the simpler EASY-PC, cannot be run inside a window – it must be used full-screen. You can freeze EASY-PC Professional screens in a window, but not make use of them. If you run all of your software from Windows, install EASY-PC Professional as a full-screen application.

Installation

If you already have the simpler EASY-PC installed on your hard drive and you have a set of files of the PCB and/or SCH type which have been made using EASY-PC, you must be particularly careful when you install the Professional version. The reason is that when EASY-PC Professional reads a file that was saved by the simple EASY-PC, it will convert the file to Professional format, and it will no longer be readable by the simpler EASY-PC. The original file is, however, saved using the P_B extension so that you can recover it if need be.

- You should always use the Standard installation unless you have been instructed otherwise by Number One Systems.

There is no problem if you have never used EASY-PC before, or if you have saved all of your drawing and layout files on floppy disks, as you should have. What you need to avoid is installing EASY-PC Professional into the same directory as you used for the simple EASY-PC package. This is normally no problem, because the Professional version will be installed into a directory called EPCPRO (the simpler version will probably be in a directory called EASYPC).

You should avoid reading any files from EASY-PC using the Professional version and then saving them back to the same EASYPC directory. The manual recommends that if you want to use these files you copy them, using the MS-DOS COPY command or Windows File Manager, into the EPCPRO directory and use them from that directory.

- Note that you cannot copy the program files from the floppy distribution disks into a hard drive directory and expect to be able to use them. The files are in compressed form and must be expanded. This is done by the INSTALL program. Some five Mbytes of hard drive space is needed.

Installation is done as follows:

1 Place Disk 1 in the A drive and type the command `A:\INSTALL`. If you are using Windows, switch to Program Manager, click on File – Run and type `A:\INSTALL`. You will see a copyright notice with the revision number of the program.

- Preferably you should install from DOS (unless you are using Windows-95). Do *not* attempt to install from DOSSHELL if you use it.

2 You are asked for a target directory. The default is C:\EPCPRO and you should use this – if you alter the directory name further manual configuration will be needed.

3 You will be asked if you want to create the directory whose name is to be used. You should accept this as a default also.

4 The files, which are in compressed format, are then copied. You will be asked (screen message and sound beep) to change to the second floppy disk when required.

5 You will be asked to return to the first distribution disk again to enter your name as a registered user. You are then asked to enter your company – this question can be answered just by pressing RETURN if you are using EASY-PC Professional in a private capacity. You are then asked to enter Town and Country. More files are then copied over.

6 When installation is complete you are informed by a message on the screen, and you can press any key to leave the installation process. This will return you to DOS or Windows, whichever you used to perform the installation.

7 Now back up your distribution disks, label the backups, and place the original disks in a safe place.

The screen display

EASY-PC Professional, when you run it for the first time, starts with a welcome screen, and you are asked to press the ↑F10 key to start work, and confirm this by pressing the Y key. The first menu provides the choice of Schematic (A), PCB (B), SCH Component (C) or PCB Component (D), SCH Symbol (E) or PCB Symbol (F), or Exit to DOS (or Windows) (key X).

- Note that if you have previously used EASY-PC Professional and you were working on a drawing or layout, this last-used drawing will appear when you next start EASY-PC Professional – the Welcome screen is not used again. If you do not want the last-used drawing to appear you can start EASY-PC Professional with the command line:

  ```
  C:\EPCPRO\EPCPRO /L-
  ```

  (Don't forget the minus sign.)

- Note also that EASY-PC Professional shows a grid even on the highest zoom number.

Figure 7.1 shows the main PCB drawing screen with four of its five menus drawn down (this is a composite view – you cannot see all of them at once). There is also an alternative first menu which is used when you have a schematic loaded and wish to use the simulator software, see later. The

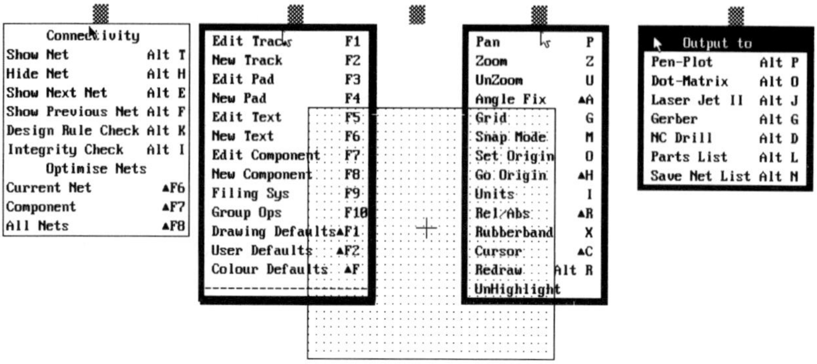

Figure 7.1 The main drawing screen of EASY-PC Professional with the four main menus displayed (composite screenshot). There is no central menu until an editing or filing action is used.

menus appear *only* when the mouse button is *clicked* (or the RETURN key pressed) – they do not appear when the cursor is over the box as was the case for EASY-PC in its simple version.

The inner three menus correspond closely with those of EASY-PC, but the outer menus offer mainly the enhanced features of EASY-PC Professional. The central menu, like that of EASY-PC, is not used until an action is selected, but on the Professional version there is a message to that effect.

Taking the objects menu first – meaning the one that deals with the main function keys and coverts insertion and editing of Tracks, Pads, Text and Components – the main changes are as follows:

1 The word *symbol* is not used for PCB layout or schematic drawing; *component* is used instead. *Symbol* now refers to shapes, *component* for the combination of shape and lettering that identifies a component. For example, a flip-flop *symbol* is common to *components* such as the 7476, 7473 and 7478.

   - Note that this makes the symbols of the Professional version incompatible with those of the Simple version – the library files *can not* be interchanged.

2 The menu contains Group Operations, which is the equivalent of Block Operations in EASY-PC. There is one considerable enhancement which allows you to add items to the group or remove items from it. This gets around the difficulty of using a strictly rectangular block. The Include key is ↑I and the Exclude key is E. In addition, the origin of a group is fixed as its bottom left-hand corner.

- You can select part of a schematic in a group to analyse the operation.

3 There is a Colour Defaults screen (↑F key) which allows you to choose colours for up to sixteen layers. The layers are numbered 0 to 9 and lettered A to F. You can designate colours and even *patterns* for Track, Pad, Text, Symbol Track, Symbol Pad, Symbol Text and Highlight.

The Control menu, which was the right-hand side menu of EASY-PC, is now the fourth from the left in EASY-PC Professional. It is almost identical to the older version, but with the addition of the UnHighlight command, key Alt-R.

The two new menus are concerned with Connectivity and Output, or alternatively, Simulation and Output.

The Connectivity menu deals with the automatic conversion of a schematic into a PCB layout. *Connectivity* means that connections that are shown in a schematic will be maintained at all times when the PCB layout is created from the schematic. The word *Net* is used to mean a connection in the schematic, so that the items in the connectivity menu that refer to Net are concerned with actions such as showing one or more connections to make inspection easier.

The Output menu is an improved way of sending a drawing or layout to a printer or plotter, drilling machine or file. The simpler EASY-PC program made use of separate programs for all output except to a dot-matrix printer; EASY-PC Professional can provide all these outputs without leaving the main program.

- Note that these additional menus make use of the ALT key along with letter keys for the short-cut key methods.

Loading a file

When you use the F9 key on EASY-PC Professional, the Browse item appears in place of the directory listing of EASY-PC. When you use Browse, it lists all the files of that type (PCB or SCH) and allows you to select one with the mouse (or cursor keys and RETURN key).

- Note that the cursor keys in EASY-PC Professional are much less sensitive than in simple EASY-PC – you can hold them down without having the cursor vanish off the screen in a fraction of a second, and it is easier to make small adjustments to the cursor position.

When a file is being loaded, you will see the status report building up, in terms of numbers of tracks, pads, components and text objects, as the file loads.

Note that the zoom levels now range from 1 to 8, one more level than EASY-PC because EASY-PC Professional allows for larger PC boards up to 32" x 32" and resolution down to 0.001".

Defaults and Status

The new User Defaults screen is shown in Figure 7.2. The facilities are the same as in the simple EASY-PC version (with the addition of selecting items to be displayed), but the method of selection is clearer and simpler. The Drawing Defaults screen now allows for sixteen numbers to refer to the different track and pad widths, and text line widths have been added to this set, Figure 7.3.

- These figures can be edited rather than just toggled or selected from the list, so that you can create defaults that are entirely your own.

The Status display is obtained, as before, with ↑F4, and now contains more information, Figure 7.4. The Reclaim Memory item can be used if in the course of your work you have deleted components. By clicking the mouse on the Reclaim Memory line you can recover the memory that was used to store these items.

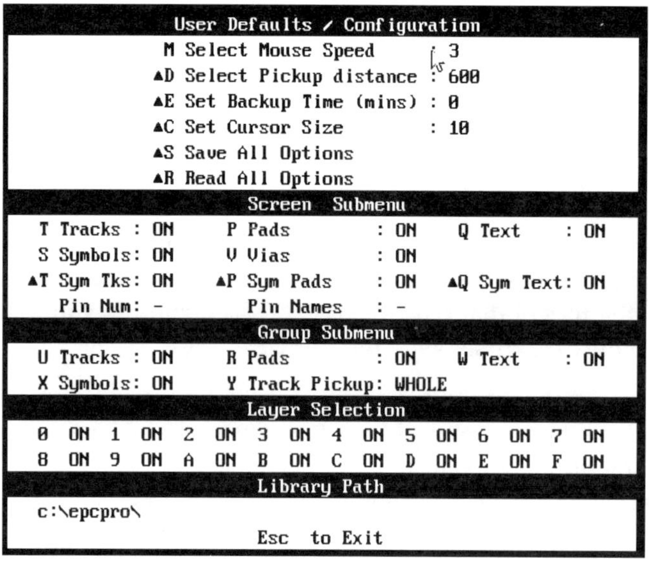

Figure 7.2 The User Defaults screen of EASY-PC Professional, showing its clearer division into sections.

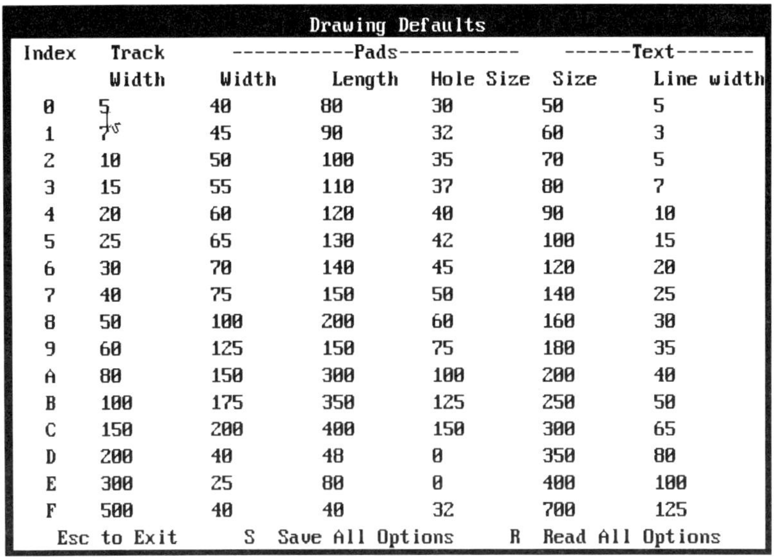

Drawing Defaults

Index	Track Width	Pads Width	Pads Length	Pads Hole Size	Text Size	Text Line width
0	5	40	80	30	50	5
1	7	45	90	32	60	3
2	10	50	100	35	70	5
3	15	55	110	37	80	7
4	20	60	120	40	90	10
5	25	65	130	42	100	15
6	30	70	140	45	120	20
7	40	75	150	50	140	25
8	50	100	200	60	160	30
9	60	125	150	75	180	35
A	80	150	300	100	200	40
B	100	175	350	125	250	50
C	150	200	400	150	300	65
D	200	40	48	0	350	80
E	300	25	80	0	400	100
F	500	40	40	32	700	125

Esc to Exit   S Save All Options   R Read All Options

Figure 7.3 The Drawing defaults for EASY-PC Professional, showing track, pad and text sizes. A set of sixteen sizes can be used for any of these objects.

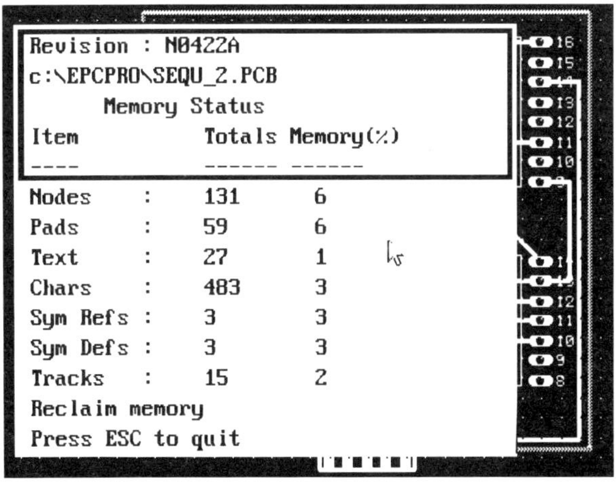

Figure 7.4 The Status Display for EASY-PC Professional with its Reclaim Memory option.

Edit menus

When objects in a drawing are edited, using the F1, F3, F5 or F7 keys, a new menu and set of key options appears in the central space. For example. the Track Edit menu is now split into two parts. The upper part contains items that are familiar from EASY-PC, but with the addition of Net Name (↑N). You can use this to find what name has been given to the connection –

EASY-PC Professional names all connections using an eight-character name like a filename. This net name appears at the bottom of the screen. The default naming uses the letter N and a number, but you can alter this, using the ↑N key, so that connections of particular interest can carry names such as CLOCK, RESET, QA, OUTPUT and so on.

The new addition to the Track Edit menu is Unroute, consisting of Part Track (↑P), Whole Track (↑FQ), or Whole Net (↑R). Unrouting is used when you have decided that your route was not so useful after all, and you want to revert to the rat's nest direct connection so that you can try another route.

The Pad editing menu now consists of Current (C), Shape (S), Angle (A), Flip (F), Length (L), Hole size (H), Width (W), Repeat (R), Delete (D) and Layer (↑L). The Angle command allows you to rotate a pad, but the result is only noticeable if the pad is asymmetrical.

- Pads that are part of a component cannot be directly edited, but you can change the symbol in the library to one with updated pad size or shape.

The Text edit display is the same as that of the simpler version, but Component Edit, replacing Symbol Edit, now contains Place. Ref. (↑P) and Update (↑U) as well as Angle (A), Flip (F), Current (C), Repeat (R) and Delete (D). The Placement Reference is the identifying name for the components, such as IC1, R3, C5, Tr7 and so on. This is used in conjunction with the schematic – the references are transferred when a schematic is converted into a PCB layout. Click on Place. Ref. or use ↑P and you will see the Placement Reference notice appear, Figure 7.5.

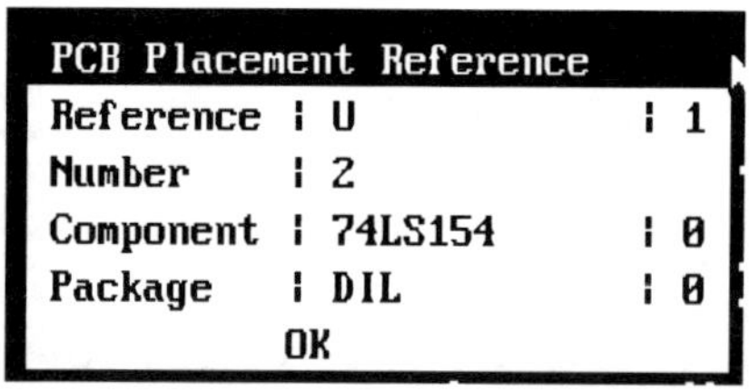

Figure 7.5 The PCB placement notice. This shows the reference (U2 in this example) for a component and its identification.

- If a Placement Reference includes a `Values: more` item at the end, clicking on this line will show an additional list. The Values list is used to contain component values for use with the Analyser software.

The Update option is intended to deal with changes in the library of components or the use of another library. If such a change is made (which is not likely for well-established components) the initial change is only to the library, and the components that have been used in drawings are not affected. Using Update on a drawing allows you to replace the existing component with the most recent version of that component. Provision is made for checking the effects of the change by way of other menus.

- Flip now moves the silk-screen of a component from layer 0 to layer F, and any simple layer pads (as used for surface mounting) from layer 1 to layer E.

Drawing a schematic

There are several examples of drawings and layouts in the EASY-PC Professional package, so that the example here is deliberately different – a four-stage Johnson (twisted-ring) counter. This has been used because of the peculiarly convoluted connections that such counters use. The main component is the 74LS73 dual J-K flip-flop, and two of these components will be needed to implement the counter. You will start, of course, by selecting the Schematic option from the main menu.

The first step is to load in the two 74LS73 Components. Use the F9 key for New Component and click on Browse. This initially offers a selection of libraries, Figure 7.6. In this set, the library of interest is called 74LS.IDX and when you click on this one you will see the first page of 74LS components, Figure 7.7.

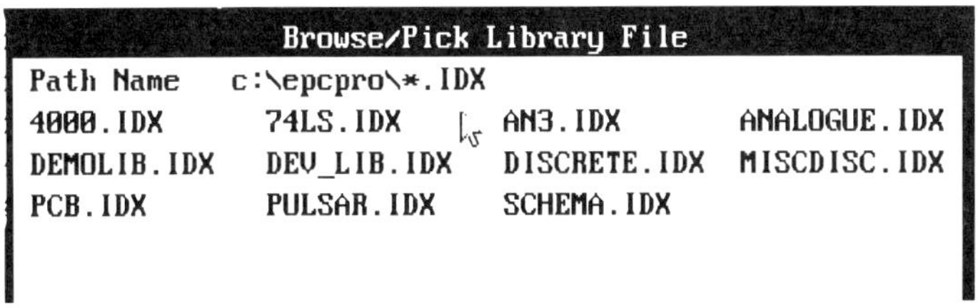

Figure 7.6 Using Browse on the library files. The main libraries of components are the 74LS series and the 4000 series.

These are in order, so that the first page contains mainly the 74LS1xx and 74LS2xx numbers – the components whose numbers start with 3 or higher are on the following page (click on *Page Down*).

When you move to the second page and click on the 74LS73 entry you will first see the Placement notice, and clicking on the OK for this will show the two separate J-K units that make up this component, with the labelling to identify the different parts, Figure 7.8. The two parts are labelled as U1a

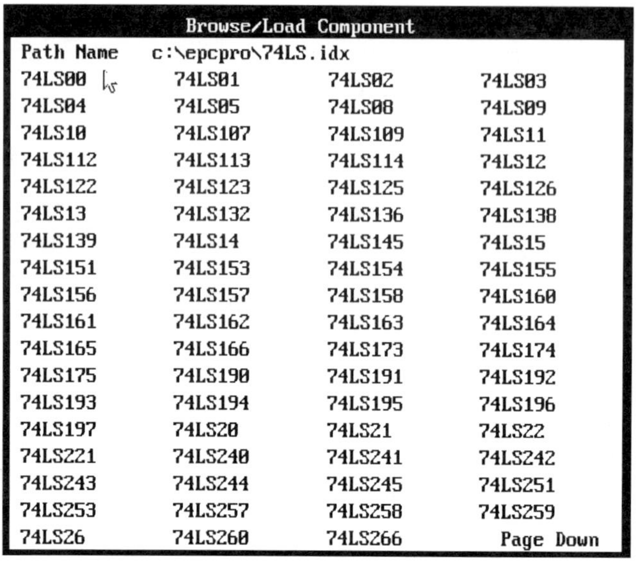

Figure 7.7 The 74LS series library opened, showing the first page of device listings. These are in order of the first numeral following the LS letters.

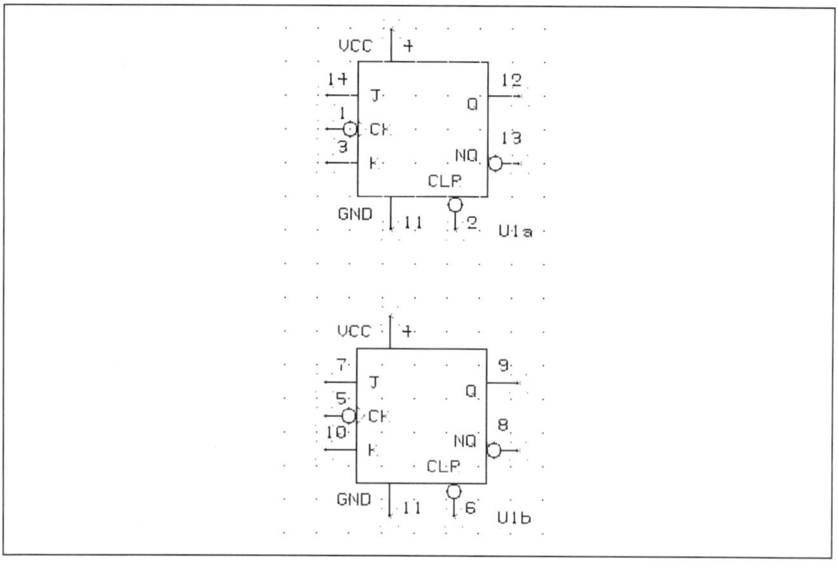

Figure 7.8 The schematic version of the 74LS73 flip-flops. The two sections are labelled U1a and U1b.

and U1b. You then repeat the component load – but this time the Placement reference box will already have another copy of the 74LS73 ready for you, labelled as U2a and U2b. The Repeat (R) action can also be used.

- The Block Repeat command is available (for compatibility with the simple EASY-PC) but it should *not* be used if you intend to employ automatic PCB creation. Note that Components are shown with a stub of connecting line that ends in a cross. This makes it easier to see where components are connected to circuit lines, and the same scheme is used for PCB components and tracks. These crosses appear only on the screen, they are not printed. Note also that circles appear correctly on the VGA screen.

You can then arrange the units and complete the schematic drawing in the way that has been described for EASY-PC. EASY-PC Professional allows you an additional Snap setting of one tenth of the grid, but this is too fine for most schematic drawing work, and you should use half or the default quarter-grid. The 0.1 snap is intended for surface-mounting PCB work.

The 74LS73 component has its origin at the J connection on pin 7 or pin 14, and you can arrange the components in line by repeating the procedure of placing the cursor, pressing F7 (Component Edit), key C for Current, and then moving the cursor to where you want the origin to be and clicking. After the first edit, moving the cursor to the second J-K flip-flop and pressing RETURN (or the left-hand mouse button) will also select it.

Making connections

The principles of making connections, or *nets*, are that lines must be drawn from node to node. You will hear an audible confirmation of a genuine connection. When you join two lines (usually at 90°) you will be prompted to confirm that it is your intention that these nets are joined.

- Note that if you place components so that their leads touch, EASY-PC Professional does *not* consider these points as being connected. You must draw a connecting track between any points that are to be connected.

Start with the clock leads.

1 Put the cursor on the first clock terminal and press the F2 (New Track) key. The loudspeaker will beep to remind you that this is the start of a line from a component pin (it beeps also when a line ends at a component pin). Draw the line across, click to put in a node, and then down.

- The beep is important. If you do not hear a beep, then no connection has been made. This is particularly important if you are using a snap setting that is considerably shorter than the grid spacing. You should never use the Freehand setting for connections.

2 Repeat this for the other four clock connections.

3 Now draw the line that joins the ends of the clock leads together and to the clock input. You will see a `Join Nets?` notice, Figure 7.9. This is used whenever lines (nets) might cross and it reminds you that in this case you want these lines electrically joined. Click on OK to accept, or press the K key. You must click on the *end* of each line to force a connection.

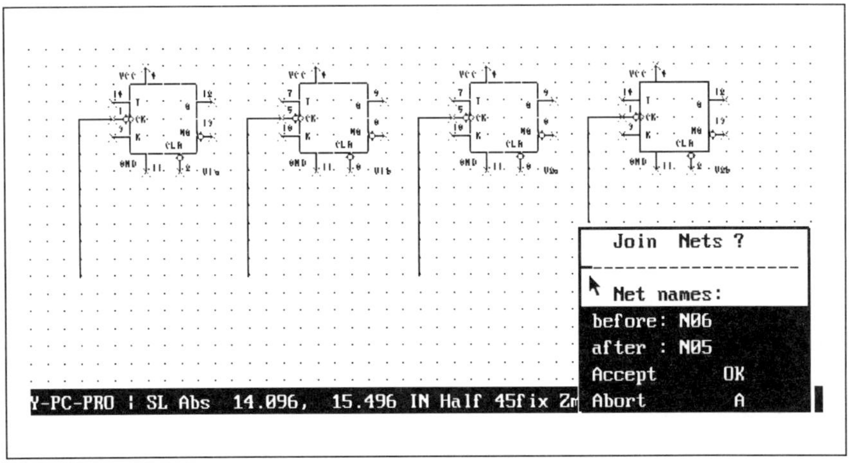

Figure 7.9 The four circuit components laid out with the clock lines. Joining the clock lines produces the 'Join Nets?' message.

- If lines cross, with no node placed at the crossing point, EASY-PC Professional will assume that no connection is intended and you will not be asked to confirm joining nets.
- Note that pressing the RETURN key does not accept the net junction.
- The rule is that a T-junction is a join, a cross is not. A line can also connect to the end of another line.

4 Complete the other pin connections, Figure 7.10.

5 The circuit will need seven connections to the PCB – supply positive and earth, clock input, and the four Q outputs. You can, if you like, use New Component to put in a D9M connector and wire up to the

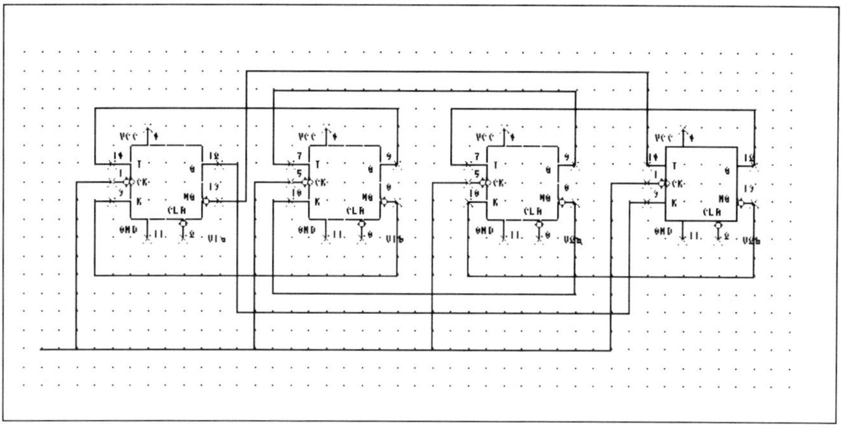

Figure 7.10 The schematic with all of its signal paths drawn in.

connection points, but this can make the PCB routing much more difficult. In this example, it is better to leave these connections hanging – they can be connected to a set of pads on the PCB layout later. Note that the CLR pins are not connected in this particular circuit – normally they would be connected together and to another pin on the D9M. You will see later that this non-connection is reported.

6 If you use the F1 key you can get a report on the Net Name (key ↑N) for the connecting line on which the cursor is placed. Names such as N07 are used, but you can type your own in place of such names. Type for the appropriate lines the names Clock, Q0, Q1, Q2 and Q3. These names will be used consistently – where lines join, the whole line will take the same name.

- Use names that consist of letters and numerals only – the use of other symbols is likely to cause problems, and it is easier to avoid symbols than to remember which can be used.

You must also be careful to check that all the different parts of a net like the Clock line are named – if any of them carries its old name it will be left unconnected when the PCB layout is created. It is usually easier to change a net name when the first line is drawn, then the other lines are automatically renamed.

7 Now save the schematic. It will at present have the default name of NEWCIRC.SCH, but you should use your own name, such as JOHNSON.SCH.

- Each time you save a schematic like this, a net-list file, with extension letters NET, is created which keeps track of all the connections that exist. This file is used by the simulation software, and is also an important part of checking that a PCB layout corresponds correctly to a schematic. You can print out this net list as a check that you have drawn the schematic correctly.

As always, you should check any schematic carefully, particularly for unconnected pins, misrouted lines or lines that do not end in a connector of some kind (such as the D9M if you used it). You can also take a look at the Status report, which for this example shows 200 nodes, 17 pads, 36 text items, 735 characters, 5 symbol references, 2 symbol definitions and 57 tracks. The memory percentages quoted show the percentage of computer memory occupied and if these are high, save and re-load the file so as to trigger the auto-allocation feature of EASY-PC Professional.

- If you work with very large layouts, there is an enhanced version of EASY-PC Professional (the XM version) which can make use of the extended memory of modern computers. Details are obtainable from Number One Systems.

The schematic can now be printed and EASY-PC Professional allows you to do this, on any type of printer that EASY-PC Professional can use, without leaving the program. As for the simpler EASY-PC, it is an advantage to draw a frame around the drawing to indicate the maximum extent of printing. You might also want to rotate the drawing to fit the paper in landscape (long-ways) rather than portrait orientation.

- Making a group will also ensure that everything in the group is printed.

Do not worry about the time that the screen remains blank when you switch to printing – this is needed to switch to the printing routine. A temporary copy of the file is also made. The printing to a laser printer is faster than that of the simple EASY-PC, and the appearance of the schematic is much better. The set of options is the same as is used in the laser printing accessory of the simple EASY-PC.

Translating to PCB Layout

We shall now convert the schematic into a PCB layout using the automatic translation feature of EASY-PC Professional.

1 With the schematic still on screen, click the extreme left-hand menu box. This contains three items: Pulsar Simulation, Analyser Simulation and Translate to PCB (ALT-C keys).

- The Simulation options are *fully* available only if you have Pulsar or Analyser (or both) installed on your computer and in the correct directories. These programs will simulate circuit action so that you can obtain a frequency and phase response graph for a linear circuit, or a pulse timing and glitch display for a digital circuit.
- EASY-PC Professional contains limited versions of Analyser and Pulsar, so that for circuits with a limited range of components you can carry out a full analysis (see later).

2 Press the ALT-C keys together to start the translation. For a small circuit like this, the translation is rapid, and you will see the zoom level change to level 8, with a PCB layout in the lower left-hand corner of the board. Zoom and pan this into the centre of the screen, Figure 7.11.

- All of the connecting lines are initially shown in thin white lines and will change to red (unless you have altered colour defaults) when you start to edit tracks.

Note that EASY-PC Professional staggers the positions of the chips so that you can see the joining lines separated. Some lines, however, run along the line of the pins and it is difficult to see at this point where they are connected.

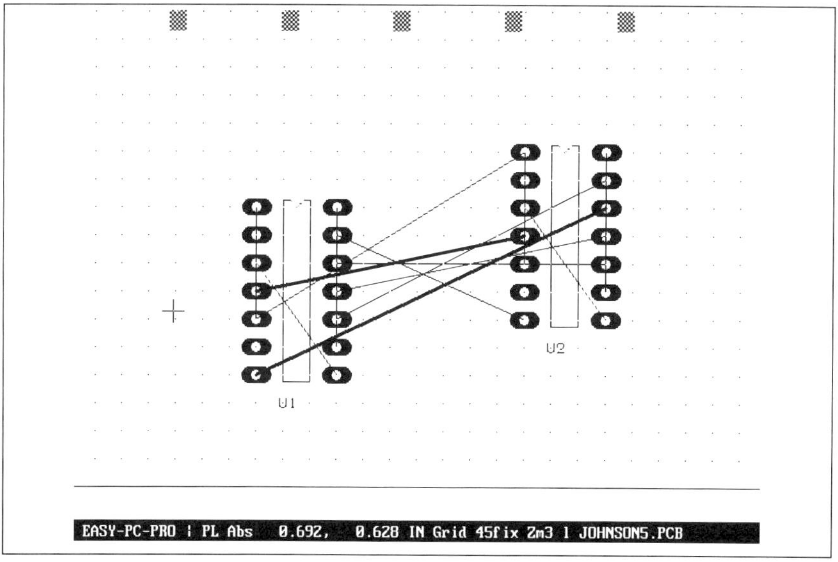

Figure 7.11 The result of translation into PCB layout. The lines have been drawn by taking the circuit connections one by one, and some could be better placed. The thicker lines show where the editing key has been used.

3 The connections that appear are the shortest straight-line connections, and you need to re-route them. You can work with either a single-layer or a multiple layer of connections, as you please.

- The most awkward lines to interpret are these which join pins on the same IC because you cannot be certain which pins they are supposed to join. You can check this from your schematic, but it is easier to edit and watch the line change colour.

4 It often helps if you shift an unrouted component around to see if the joining lines make fewer crossings at different positions of the component. If this leaves the drawing looking patchy, use the P key with the cursor in the centre of the drawing; or use the ALT-R keys for Redraw (without changing the screen). It is also helpful to switch off the grid temporarily.

- Remember also that if you designed the schematic using a connector you will probably have used the most convenient connector pins as they appear, in a straight line, on the circuit. Do not feel committed by these – you can swap pin use around at the PCB stage as long as you note how the pins are used and amend the schematic also.
- EASY-PC Professional will maintain connectivity as long as the schematic and the PCB are not changed independently.

5 EASY-PC Professional can also help, however. If you look at the menu on the extreme left-hand side, you will see that there is an *Optimise Nets* section. This is used to perform automatic re-routing in a straight line and, for this example, the best choice is *All Nets*. This will ensure that when you have moved components around, the best straight-line connections will be re-drawn. Figure 7.12 shows this done for the example.

- Note that this does *not* re-route the lines so that they are ready for use, it only finds the most direct straight-line path where strict adherence to the schematic connections has created more tortuous paths. You can get a report on the number of crossed lines from the Design Rule option.
- You can also use the Integrity Check to see if all the connections are correct. If both the schematic and the PCB layout have been saved, this will have created net files for both. If neither has been saved (which is very bad practice) you can use the Save Net List option on the fifth menu. The Integrity Check will show Pass or Fail when it is complete, and if the message is Fail, files which carry

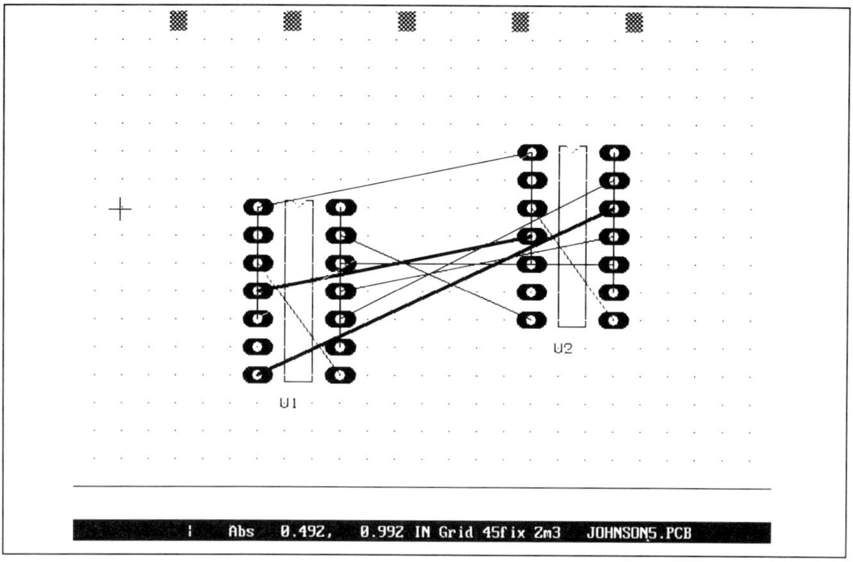

Figure 7.12 Using the All Nets option of Optimise Nets produces this re-routing.

the PDN and SDN extensions, for example MYDRAW.PDN and MYDRAW.SDN, will be created.

- These files will not be created if the result is a Pass. Once you see the Pass or Fail message you can continue work. The integrity check operates by comparing the net files, which is why it is important to save schematics and layouts so as to create these net files.

6 You can pull out the lines that are confusing, so that you can see what pins they should connect to, Figure 7.13. Following this you should check that Track Pickup is set to Partial, and use a Group Move to shift the components into line, Figure 7.14.

- Note that the group, unlike a block of EASY-PC, has its origin in the bottom left-hand corner, irrespective of where you started drawing the group.
- In this example, the number of crossed tracks makes it appear that it might be an advantage to rotate one of the chips by 180°. This makes several tracks almost straight, and you can now go on to edit the remainder. This expedient is reasonable if you are making a one-off PCB for yourself, but for commercial production all IC chips must face the same way.

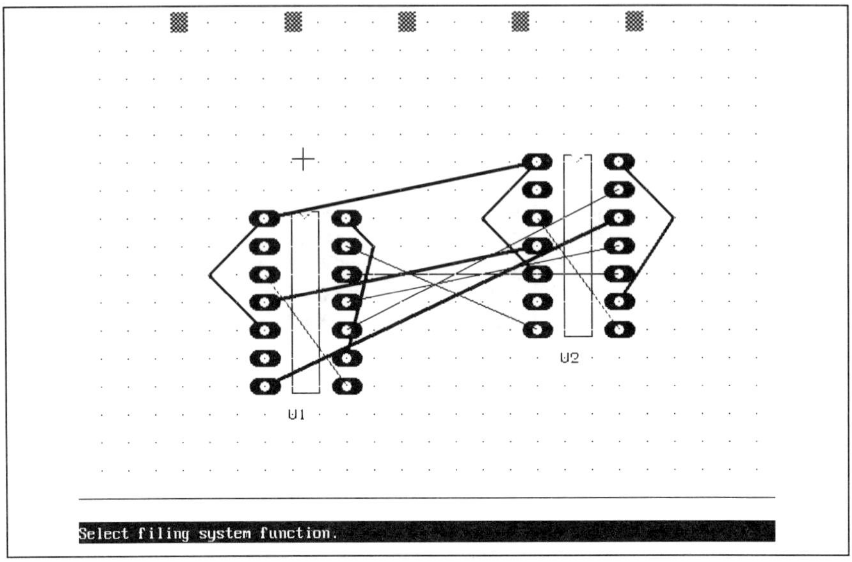

Figure 7.13 A first effort in re-routing. The lines that run between pins on the same side of the chip have been pulled out to make it clear how they are connected.

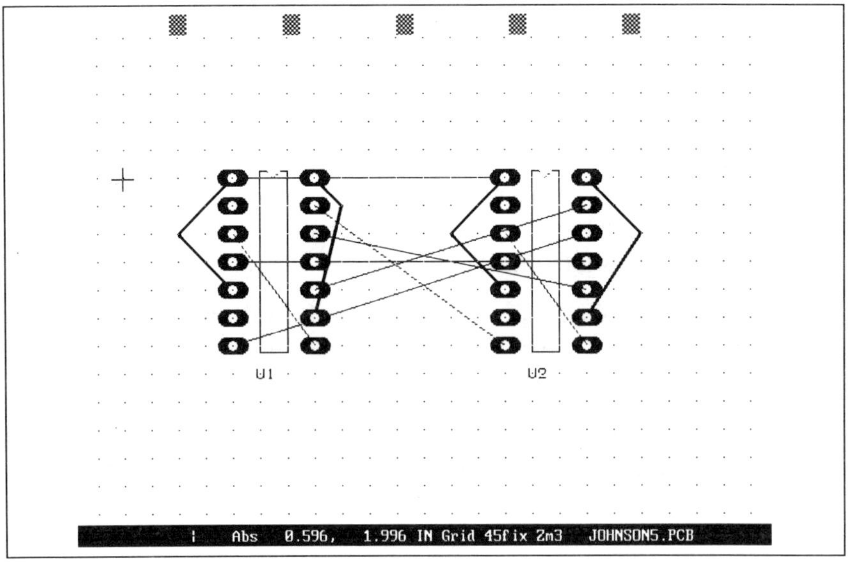

Figure 7.14 The chips put into line by using a Group Shift command. This must be done before the final track editing can be carried out.

When you have edited the tracks to what you think is a good re-routing, you should use the Design Rule check to ensure that no crossed tracks remain. Do not accept a layout until the Design Rule report shows that there are no remaining errors.

Figure 7.15 shows another step to a final layout with tracks edited so as not to cross. The only remaining problem is to make connections to the Q outputs on pins 9 and 12 of each IC. For an example of this size, and for small-scale production, it is probably easier to use wire links to pads for some of these outputs.

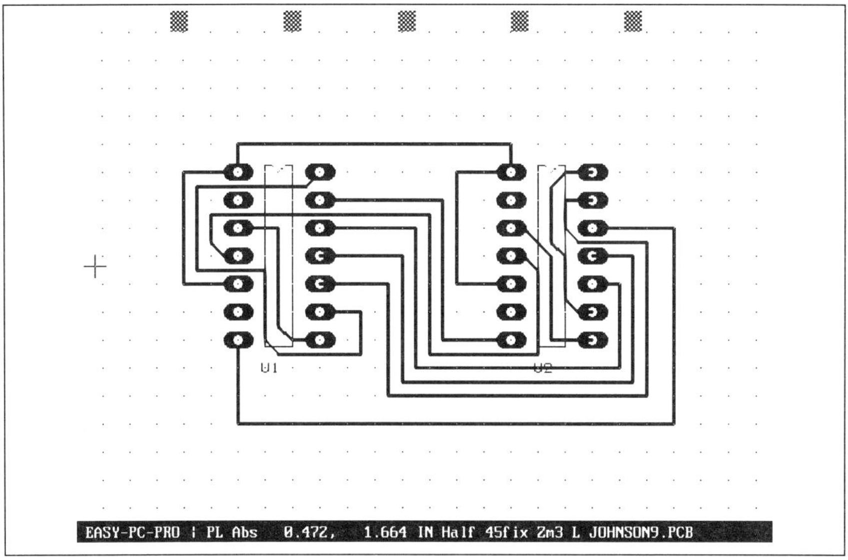

Figure 7.15 The next step to a final layout with tracks edited so as to avoid all crossings. The outputs have not been taken from the Q pins yet, and in the final version you would need to use links to pads for these outputs.

If you have any doubts about whether the PCB layout is identical to the schematic, you can use the Integrity Check – remember that you must make sure that the net file has been created (click on Save Net List in the fifth menu if you have not saved the schematic or PCB files). The net list for the schematic must also have been created, because the integrity check operates by comparing the lists. If either PCB or schematic has its connections altered, a new net list must be created.

- The integrity check should be used only where the design either fits on a single-sided board without links or on a double-sided board without links. However, if you set the via size to match the pad size for the links and create the links on another layer (as if on a double-sided board) the integrity check can still be used.

- Note that in EASY-PC Professional, using multi-layer board, a via *must* be made automatically – there is no provision for inserting a via manually by using width 16 as there was in the simpler EASY-PC.

You need to exercise some patience until you are fully accustomed to EASY-PC Professional, because it inevitably takes some time to become accustomed to track editing unless you have already trained yourself with EASY-PC. You will, however, soon become familiar with the methods that EASY-PC Professional uses, and you will soon appreciate the importance of maintaining connectivity.

Appendix A: Gerber photo-plot output

The files produced by EASY-PC can be converted into the format needed by the Gerber photo-plotting machine. These machines are operated by photo-plotting bureaux and by other specialists in PCB production, and if you have never made use of their services you are not likely to be familiar with the methods. The older plotting machines made use of a disc which was drilled with a set of different apertures. By moving and rotating this disk, light passing through one selected aperture on to a sensitized film would expose patterns which would become pads, tracks and copper lettering on the final PCB.

Later types of plotter have used laser beams in place of the aperture disc, but both types follow the same methods of computer control which rely on a set of instructions on a computer disk. The laser type is now preferred because it can be more completely controlled by software – the apertures, for example, are not preset by a disc. In either case, the output of the Gerber machine is a transparent film that can be used for PCB production using any of the current methods. Many of the bureaux that offer photo-plotting services will also offer PCB production as well.

If you want to send a PCB layout to a bureau, the first step is to find a local bureau and make contact. Number One Systems can provide a list of bureaux and PCB manufacturers that will provide these services, along with the name of someone to contact in each bureau. You should discuss your requirements in full to ensure that your design can be produced and, if so, can be produced economically. You may need to modify your design slightly, and it helps greatly if you can produce a good plot on paper as an illustration, along with a list of track and pad widths.

Producing the Gerber files requires the use of the EASYGERB utility, which should have been installed along with EASY-PC if you intended to use this facility. You will also need sufficient formatted floppy disks to take the large files that even a fairly simple layout will generate. The floppy disks must be formatted ready for use.

1 Save your layout file in PCB form, for example MYPLOT.PCB.

2 Leave EASY-PC and start the EASYGERB program, typing the filename separated by a space. For example, `EASYGERB MYPLOT` would be used to create the files for the MYPLOT.PCB file.

3 The screen will display an analysis of the circuit in terms of layers, resist masks and other options. Figure A.1 shows this for the EASY-PC Professional version of the Gerber output.

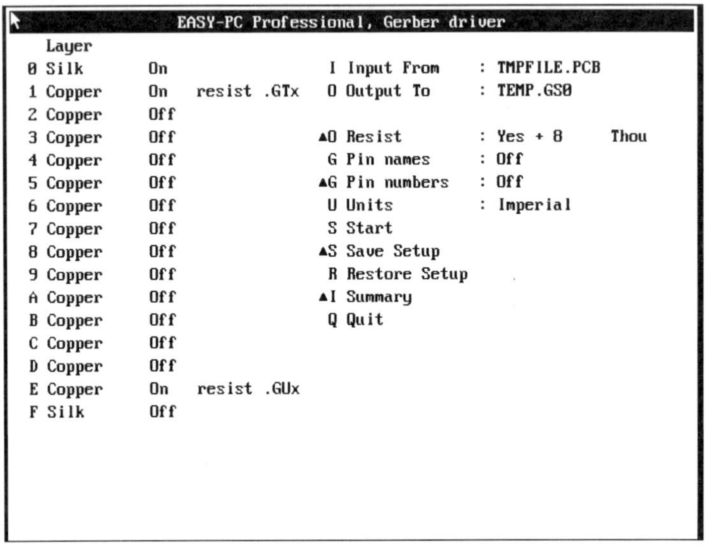

Figure A.1 The Gerber driver main menu – this example is taken from EASY-PC Professional (using the file PROTGERB.EXE).

4 You can enable or disable whatever layers you want. Remember that layer 1 will be the output for the top solder mask pattern, and layer 9 is the output for the bottom solder mask.

5 The Resist option is on by default. The figure following the Yes for this option is the clearance, in 0.001" units, over the diameter of the hole. Normally you would allow 0.012" oversize for the solder layer on a pad.

6 You can change the input filename and the output filename as required – EASY-PC Professional uses a temporary file TMPFILE.PCB and you would not normally want to change this. If you change the output filename in EASYGERB use *only six characters* because the Gerber driver will add another two. You can specify a drive if you need to use a drive other than the A drive.

7 Click on Start Plot (in EASY-PC Professional only) or use the S key to start the plot. This will normally be to the A drive, and there must be a formatted disk in that drive. You will be prompted to change disks if required.

8 Quit the program by using the E option in EASYGERB, or the Q option in EASY-PC Professional. This returns you to DOS (or Windows, if you started from Windows) or back to your design in EASY-PC Professional.

Note that if you have Gerber files that were produced using other PCB layout programs, you can use these files to produce EASY-PC layout files. This is done using the EASYLINK program obtainable from Number One Systems.

Appendix B: Drilling machines

EASY-PC and EASY-PC Professional can send outputs to numerically-controlled drilling machines of the Excellon type so that pad holes can be drilled for a PCB. If you are using EASY-PC, you must save the file for the layout you have completed, and use the EASYDRIL program. If you are using EASY-PC Professional then the Outputs menu contains the Drilling Machine option.

The procedure for creating floppy disks for a drilling machine is almost identical to that for the Gerber process in Appendix A. You should consult the PCB manufacturer to check that the drill sizes for your PCB are in stock, and that your layout is suitable. Once again, it is useful to send a paper plot to the manufacturer. The EASYDRIL program contains an Adjust Drill Sizes option which allows the tool sizes to be changed without needing to alter your pads. This is useful if your layout would call for drills that are not in stock, or if it would necessitate too many tool changes for economical production.

Appendix C: Pulsar and Analyser-3

Making and testing prototypes of a circuit is a time-consuming process, and one which is liable to all kinds of errors. A considerable amount of time can be saved by using computer analysis of a circuit, which allows the circuit to be emulated and its behaviour plotted.

For example, Analyser-3 is used to check analogue circuits, and it will produce graphs of amplitude and phase response plotted against frequency. In addition, input and output impedance values can be plotted over the same frequency range. These graphs allow you to dispense with the use of a prototype, particularly if you can make reasonable guesses at the value of stray capacitances.

The Pulsar program performs analysis for digital circuits, allowing you to enter pulse at any repetition frequency or mark/space ratio that would be used. The outputs and intermediate waveforms can be plotted, and the program will detect a glitch, even one that occurs only once in a day or so.

These two programs are integrated with EASY-PC Professional, so that when a schematic is drawn using EASY-PC Professional it can then be analysed. If the results of the analysis are favourable, the PCB layout can then be generated, and the boards put into production with an immense saving in time and materials. Your copy of EASY-PC Professional contains versions of Analyser and Pulsar that permit these actions to be carried out on a limited scale, and an example of what is possible with this built-in capability can be illustrated by the FSK schematic. This contains both analogue and digital circuitry, illustrating that Analyser will provide results for the analogue portion, and Pulsar will provide results for the digital section.

The FSK schematic is illustrated in Figure C.1 – the zoom used here does not show the fine detail of lettering. Also in this illustration you can see the results of clicking on the left-hand menu box, revealing the Simulation options.

When Analyser is run, you will be warned that component U1 cannot be found. This is because Analyser cannot work on digital ICs, and if you press

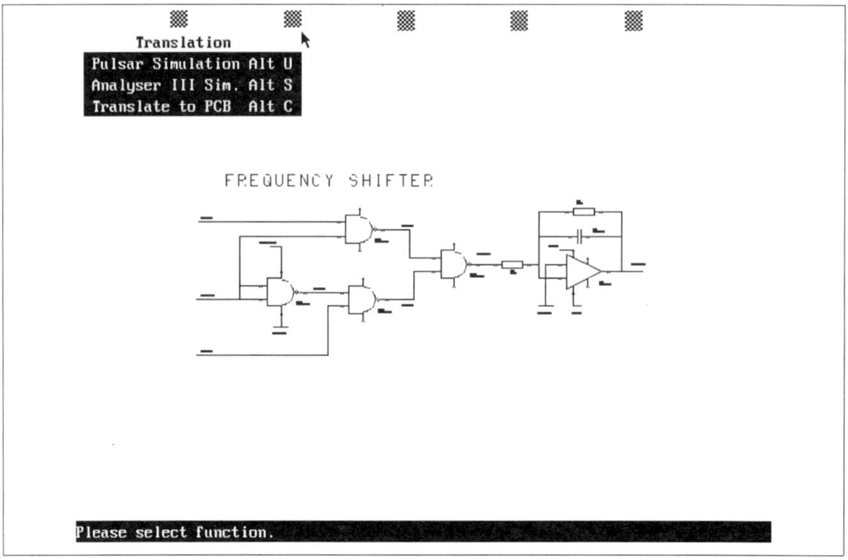

Figure C.1 The FSK circuit, one of the examples included with EASY-PC Professional. this circuit contains both digital and linear portions.

the RETURN key this component will be ignored and the linear simulation will be run on the linear part of the circuit.

When the Analyser run is complete (very quickly on this simple circuit) you will see the screen report in the form of a graph of amplitude and phase plotted against frequency, Figure C.2. The phase graph line is in red and the gain (amplitude) line is in white. Phase angle is measured in degrees with a scale on the right-hand side of the screen. Gain is measured in dB on the left-hand side of the screen, with a common frequency scale, in logarithmic steps, at the foot of the screen. You can escape from the plotted display by pressing the Q key or by clicking the mouse over the Quit option at the top of the screen. You will be asked to confirm this.

When you opt to use the Pulsar simulation for the digital part of the circuit you will be notified this time that the component U2 cannot be found. This is because Pulsar is a digital simulator and will not recognize linear components, and if you press the RETURN key when the message appears the simulation will continue.

The inputs to the FSK circuit are FREQ1, FREQ2 and SELECT, and these same names will appear in the simulation. The outputs are shown at circuit nodes, labelled as Node 1 to Node 3, and at the output labelled SELOUT. The graphical output of Pulsar is shown for this example in Figure C.3.

When you return from the Analyser or Pulsar examples you will see a screen that describes how to upgrade to the full version of each.

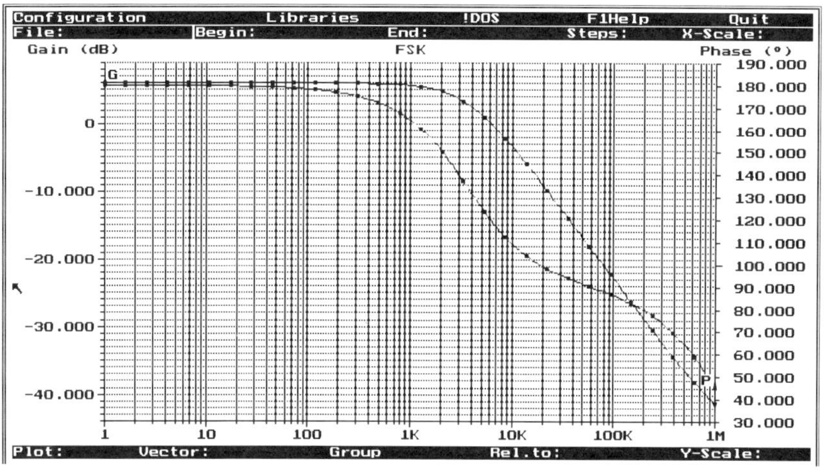

Figure C.2 The output of the Analyser simulation for the linear part of the FSK circuit, showing the graph lines of gain and phase plotted against frequency.

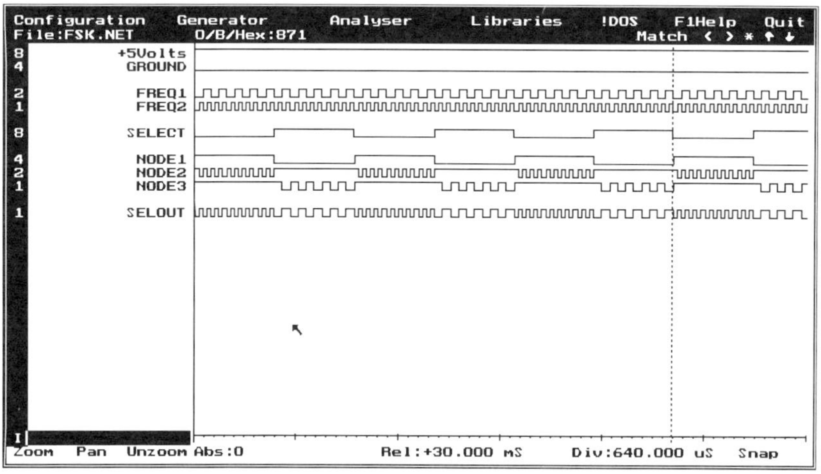

Figure C.3 The output of the Pulsar simulation for the digital portion of the FSK circuit, showing the input, intermediate and output pulse waveforms.

Appendix D: Outline of PCB preparation

Crude early models of circuits can be constructed on stripboard, but for anything other than a one-off experiment PCBs must be produced which are likely to be put into mass production later. A board material must be used which will be strong and heat resistant, with good electrical insulation. The choice will usually be between plastic-impregnated glass fibre board, or SRBP (Synthetic Resin Bonded Paper), though some special-purpose circuits may have to be laid out on ceramic (like porcelain) or vitreous (like glass) materials in order to cope with high-temperature use and flameproofing requirements. Materials other than SRBP are much more costly, and are likely to be used only when a military specification must be followed or when special thermal characteristics must be met. A special case is flexible PCBs as used in the printheads for dot-matrix or ink-jet printers.

The pattern of the circuit tracks has to be drawn on a piece of copper-laminated board. For a DIY one-off this can be done using a felt-tipped pen which contains etch resistant ink. Another option which is slower but which provides better-looking pads and tracks is to use etch-resistant transfers. Standard transfer patterns include lines of various thicknesses, i.c. and transistor pads, and pads for mounting connectors, together with a variety of curves, dots, triangles and other patterns.

For work of medium precision that requires a fairly large number of PCBs to be produced, the patterns can be applied by a silk-screen process. This is inexpensive and produces good clean work, though the thickness of track lines is not consistent enough to allow very narrow lines to be used.

The usual commercial mass-production method uses a transparency of the pattern as a master. This transparency can for some types of work be printed directly from EASY-PC using draughting or tracing paper or heat-resisting transparent film, but the usual route is by way of a PCB bureau or manufacturer (see Appendix A).

The transparency is then used in a photographic process. The copper side of the PCB is coated with a light-sensitive etch-resistive material which is exposed to light through the transparency, using much the same method as was used for contact printing in older systems of photographs (when negative sizes were so large that enlargers were unnecessary).

The exposed board is then 'developed' in a sodium hydroxide solution. When the ink or other etch-resist is dry, the board is etched in an acidified ferric chloride bath (acid hazard – wear goggles, gloves and an apron) until all of the unwanted copper has been removed. The board is washed in demineralized water and in alcohol and dried. The holes are then drilled. For a small-scale board this can be done by hand, but for mass-produced boards a computer-controlled drilling machine will be used (see Appendix B).

Circuit boards for commercial use have been of almost a stereotyped pattern until the comparatively recent rise in the use of surface-mounted components. Many suppliers offer such boards with the copper already coated with photoresist, saving considerable time and effort for small batches. These pre-coated boards must be stored carefully, preferably at low temperatures between 2°C and 13°C, and have a shelf life which is typically 1 year at 20°C. The maximum allowable temperature is 29°C.

Board sizes now follow the Eurocard standards of 100 mm x 160 mm, 100 mm x 220 mm, 233.4 mm x 160 mm and 233.4 mm x 220 mm; and there are also the older sizes of 203 mm x 95 mm (8 x 3.75 inches) and 304.8 mm x 457.2 mm (12 x 18 inches). Boards can be obtained with edge-connecting tongues already in place – these must, of course, be masked when the main board is etched. When boards are bought uncoated, photoresist can be sprayed as an aerosol for small-scale production or R&D applications.

Many commercial PCBs, particularly for computer or other digital applications, are double-sided, with tracks on the component side as well as on the conventional track side. Where connections are needed between sides, plated-through holes are used. These are holes which have copper on each side and which have been electroplated with copper so that the holes have become partly filled, making a copper contact between the sides. These connections are strengthened when the board is soldered.

NOTE: A UK source for Iso Draught Transparentiser (for making paper prints translucent) is:

Cannon and Wrinn,
68 High St.,
Chislehurst,
Kent

Tel: (0181) 467 0935

Appendix E: Suggested default values

The following tables are for suggested User Defaults for three different types of board: single-sided, double-sided with through plating and surface mount. The code numbers are shown along with dimensions in thousandths of an inch and in microns (thousandths of a millimetre).

Surface mount defaults table

Number	Track		Pad Width		Pad hole	
	thou	microns	thou	microns	thou	microns
1	0.2	5.1	10	25.4	2	5.1
2	6	5.2	15	38.1	2	5.1
3	10	25.4	19	48.3	2	5.1
4	15	38.1	27	68.6	2	5.1
5	23	58.4	40	101.6	2	5.1
6	31	78.7	52	132.1	23	58.4
7	69	175.3	56	142.2	27	68.6
8	148	375.9	60	152.4	31	78.7
9	65	165.1	31	78.7		
10	69	175.3	31	78.7		
11	81	205.7	31	78.7		
12	110	279.4	40	101.6		
13	119	302.3	55	139.7		
14	85	215.9	40	101.6		
15	215	546.1	106	269.2		
16	48	121.9	16	40.6		

Single-sided board default table

Number	Track		Pad Width		Pad hole	
	thou	microns	thou	microns	thou	microns
1	6	15.2	60	152.4	23	58.4
2	15	38.1	60	152.4	31	78.7
3	27	68.6	81	205.7	27	68.6
4	52	132.1	85	215.9	40	101.6
5	81	205.7	127	322.6	31	78.7
6	12	30.5	127	322.6	40	101.6
7	202	513.1	127	322.6	47	119.4
8	460	1168.4	127	322.6	55	139.7
9	148	375.9	69	175.3		
10	169	429.3	85	215.9		
11	202	513.1	98	248.9		
13	215	546.1	106	78.7		

Double-sided plated-through defaults table

Number	Track		Pad Width		Pad hole	
	thou	microns	thou	microns	thou	microns
1	2	5.08	60	152.4	23	58.4
2	10	25.4	60	152.4	31	78.7
3	15	38.1	81	205.7	27	68.6
4	27	68.6	81	205.7	31	78.7
5	36	91.4	85	215.9	40	101.6
6	81	205.7	110	279.4	31	78.7
7	152	386.1	110	279.4	40	101.6
8	302	767.1	127	322.6	31	78.7
9	127	322.6	40	101.6		
10	127	322.6	47	119.4		
11	127	322.6	55	139.7		
12	148	375.9	69	269.2		
15	252	640.1	127	322.6		
16	60	152.4	23	58.4		

Appendix F: Symbol libraries

The symbol libraries that come with EASY-PC cover the main TTL and 4000-series CMOS chips along with a large number of discrete components. You can buy additional symbol libraries for more specialized uses and at the time of writing the main library and the Surface Mount library disks were available.

The main library contents are listed below. Each Eurocard type 6464, 6496 and 9696 is available as Single (S), Long Single (SL), Double (D) or Long Double (DL), making 12 varieties in all. This library also contains templates for a set of boards that are used for PC adapter cards and other purposes as follows:

PC16L	PC16S	PC8S	PC8L	PC8M
PC8S	AT16L	AT16S		
EU6464S	EU6464SL	EU6464D	EU6464DL	
EU6496S	EU6496SL	EU6496D	EU6496DL	
EU9696S	EU9696SL	EU9696D	EU9696DL	

The IC chips that are catered for are as follows:

4000 Series Library – 4000.SLB				
4000UB	4001B	4001UB	4002UB	4002B
4006B	4007UB	4008	4009UB	4010B
4011UB	4011B	4012B	4013B	4014B
4015B	4016B	4017B	4018B	4019B
4020B	4021B	4022B	4023B	4023UB
4024B	4025UB	4025B	4027B	4028B
4029B	4030B	4035B	4040B	4041UB
4042B	4043B	4044B	4046B	4047B
4049B	4050B	4051B	4052B	4053B
4060B	4066B	4067B	4068B	4069UB
4070B	4071B	4072B	4073B	4075B
4076B	4077B	4078B	4081B	4082B
4086B	4093B	4094B	4099B	40106B
4174B	4175B	4194B	4502B	4503B
4508B	4510B	4511B	4512B	4513B
4514B	4515B	4516B	4517B	4518B
4519B	4520B	4521B	4522B	4526B
4527B	4528B	4531B	4532B	4538B
4539B	4541B	4543B	4555B	4556B
4557B	4584B	4585B		

74AC Series Library – 74AC.SLB				
74AC00	74AC04	74AC05	74AC08	74AC32
74AC74	74AC86	74AC109	74AC138	74AC139
74AC157	74AC158	74AC161	74AC163	74AC191
74AC193	74AC238	74AC240	74AC241	74AC244
74AC245	74AC257	74AC258	74AC280	74AC283
74AC373	74AC374	74ACS40	74AC541	74AC563

74ACT Series Library – 74ACT.SLB				
74ACT00	74ACT04	74ACT05	74ACT08	74ACT32
74ACT74	74ACT86	74ACT109	74ACT138	74ACT139
74ACT157	74ACT158	74ACT161	74ACT163	74ACT191
74ACT193	74ACT238	74ACT240	74ACT241	74ACT244
74ACT257	74ACT258	74ACT280	74ACT283	74ACT374
74ACT563				

74ALS Series Library – 74ALS.SLB				
74ALS00	74ALS02	74ALS03	74ALS04	74ALS05
74ALS09	74ALS10	74ALS11	74ALS20	74ALS21
74ALS27	74ALS30	74ALS32	74ALS33	74ALS37
74ALS38	74ALS74	74ALS86	74ALS1008	74ALS109
74ALS112	74ALS133	74ALS138	74ALS139	74ALS151
74ALS153	74ALS157	74ALS161	74ALS163	74ALS174
74ALS175	74ALS240	74ALS241	74ALS244	74ALS245
74ALS253	74ALS257	74ALS299	74ALS373	74ALS374
74ALS534	74ALS540	74ALS541	74ALS569	74ALS573
74ALS574	74ALS580	74ALS616	74ALS640	74ALS645
74ALS646	74ALS652	74ALS857	74ALS873	74ALS874

74AS Series Library – 74AS.SLB				
74AS00	74AS04	74AS08	74AS10	74AS32
74AS74	74AS244	74AS373	74AS573	74AS574
74AS652				

74BCT Series Library – 74BCT.SLB				
74BCT125	74BCT240	74BCT241	74BCT244	74BCT245
74BCT373	74BCT374	74BCT534	74BCT540	74BCT541
74BCT623	74BCT640			

74C Series Library – 74C.SLB				
74C00	74C02	74C04	74C08	74C10
74C14	74C20	74C30	74C32	74C42
74C48	74C73	74C74	74C76	74C85
74C86	74C90	74C93	74C150	74C151
74C154	74C157	74C160	74C161	74C163
74C164	74C165	74C173	74C174	74C175
74C192	74C193	74C221	74C240	74C244
74C373	74C374	74C901	74C902	74C903
74C904	74C906	74C907	74C914	74C941

74F Series Library – 74F.SLB				
74F00	74F02	74F04	74F08	74F10
74F11	74F13	74F14	74F20	74F27
74F32	74F37	74F38	74F40	74F51
74F64	74F74	74F85	74F86	74F109
74F125	74F126	74F132	74F138	74F139
74F151	74F153	74F157	74F158	74F161
74F163	74F164	74F169	74F174	74F175
74F181	74F194	74F240	74F241	74F244
74F245	74F253	74F257	74F259	74F273
74F280	74F283	74F365	74F366	74F367
74F368	74F373	74F374	74F378	74F381
74F395	74F399	74F655	74F656	74F657
74F1242	74F1243			

74FCT Series Library – 74FCT.SLB

74FCT240	74FCT241	74FCT244	74FCT245	74FCT273
74FCT373	74FCT374	74FCT540	74FCT541	74FCT563
74FCT573	74FCT574			

74HC Series Library Part 1 – 74HC(1).SLB

74HCOO	74HC02	74HC03	74HC04	74HC08
74HC10	74HC11	74HC14	74HC20	74HC27
74HC30	74HC32	74HC42	74HC51	74HC58
74HC73	74HC74	74HC75	74HC76	74HC85
74HC86	74HC107	74HC109	74HC112	74HC113
74HC123	74HC125	74HC126	74HC132	74HC133
74HC137	74HC138	74HC139	74HC147	74HC151
74HC153	74HC154	74HC157	74HC158	74HC160
74HC161	74HC162	74HC163	74HC164	74HC165
74HC173	74HC174	74HC175	74HC181	74HC182
74HC192	74HC193	74HC194	74HC195	74HC221
74HC237	74HC240	74HC241	74HC242	74HC243
74HC244	74HC245	74HC251	74HC253	74HC257
74HC259	74HC273	74HC280	74HC298	74HC299
74HC354	74HC356	74HC365	74HC366	74HC367
74HC368	74HC373	74HC374	74HC390	74HC393
74HC423	74HC533	74HC534	74HC540	74HC541
74HC563	74HC564	74HC573	74HC574	

74HC Series Library Part 2 – 74HC(2).SLB

74HC589	74HC595	74HC597	74HC640	74HC643
74HC646	74HC648	74HC688	74HC4002	74HC4016
74HC4017	74HC4020	74HC4024	74HC4040	74HC4049
74HC4050	74HC4051	74HC4052	74HC4053	74HC4060
74HC4066	74HC4075	74HC4078	74HC4316	74HC4351
74HC4352	74HC4353	74HC4511	74HC4514	74HC4538
74HC4543	74HC7266			

74HCT Series Library Part 1 – 74HCT(1).SLB				
74HCT00	74HCT02	74HCT03	74HCT04	74HCT08
74HCT10	74HCT11	74HCT14	74HCT20	74HCT21
74HCT27	74HCT30	74HCT32	74HCT42	74HCT73
74HCT74	74HCT75	74HCT85	74HCT86	74HCT93
74HCT107	74HCT109	74HCT112	74HCT123	74HCT125
74HCT126	74HCT132	74HCT137	74HCT138	74HCT139
74HCT147	74HCT151	74HCT154	74HCT157	74HCT158
74HCT160	74HCT161	74HCT162	74HCT163	74HCT164
74HCT165	74HCT166	74HCT173	74HCT174	74HCT175
74HCT181	74HCT182	74HCT190	74HCT191	74HCT192
74HCT193	74HCT194	74HCT195	74HCT221	74HCT237
74HCT238	74HCT240	74HCT241	74HCT242	74HCT243
74HCT244	74HCT245	74HCT251	74HCT253	74HCT257
74HCT258	74HCT259	74HCT273	74HCT280	74HCT283
74HCT297	74HCT299	74HCT354	74HCT356	74HCT365
74HCT366	74HCT367	74HCT368	74HCT373	74HCT374
74HCT377	74HCT390	74HCT393	74HCT423	74HCT533
74HCT534	74HCT540	74HCT541	74HCT563	

74HCT Series Library Part 2 – 74HCT(2).SLB				
74HCT564	74HCT573	74HCT574	74HCT597	74HCT640
74HCT643	74HCT646	74HCT648	74HCT670	74HCT688
74HCT4002	74HCT4015	74HCT4016	74HCT4017	74HCT4020
74HCT4024	74HCT4040	74HCT4051	74HCT4052	74HCT4053
74HCT4060	74HCT4066	74HCT4075	74HCT4094	74HCT4316
74HCT4351	74HCT4352	74HCT4353	74HCT4511	74HCT4514
74HCT4538	74HCT4543	74HCT7046		

74LS Series Library Part 1 – 74LS(1).SLB				
74LS00	74LS01	74LS02	74LS03	74LS04
74LS05	74LS06	74LS07	74LS08	74LS09
74LS10	74LS11	74LS12	74LS13	74LS14
74LS15	74LS20	74LS21	74LS22	74LS26
74LS27	74LS28	74LS30	74LS32	74LS33
74LS37	74LS38	74LS40	74LS42	74LS47
74LS48	74LS49	74LS51	74LS54	74LS55
74LS73	74LS74	74LS75	74LS76	74LS77
74LS78	74LS83	74LS85	74LS86	74LS90
74LS91	74LS92	74LS93	74LS95	74LS107
74LS109	74LS112	74LS113	74LS114	74LS122
74LS123	74LS125	74LS126	74LS132	74LS133
74LS136	74LS137	74LS138	74LS139	74LS145
74LS147	74LS148	74LS151	74LS153	74LS154
74LS155	74LS156	74LS157	74LS158	74LS160
74LS161	74LS162	74LS163	74LS164	74LS165
74LS166	74LS168	74LS169	74LS170	74LS173
74LS174	74LS175	74LS181	74LS182	74LS183

74LS Series Library Part 2 – 74LS(2).SLB				
74LS190	74LS191	74LS192	74LS193	74LS194
74LS195	74LS196	74LS197	74LS221	74LS240
74LS241	74LS242	74LS243	74LS244	74LS245
74LS247	74LS248	74LS249	74LS251	74LS253
74LS256	74LS257	74LS258	74LS259	74LS260
74LS266	74LS273	74LS279	74LS280	74LS283
74LS290	74LS293	74LS295	74LS298	74LS299
74LS321	74LS323	74LS348	74LS352	74LS353
74LS354	74LS356	74LS365	74LS366	74LS367
74LS368	74LS373	74LS374	74LS375	74LS377
74LS385	74LS386	74LS390	74LS393	74LS395
74LS398	74LS399	74LS442	74LS444	74LS490
74LS540	74LS541	74LS568	74LS569	74LS590
74LS592	74LS593	74LS595	74LS597	74LS604
74LS605	74LS606	74LS607	74LS612	74LS620
74LS621	74LS622	74LS623	74LS625	74LS630
74LS640	74LS641	74LS642	74LS643	74LS644
74LS645				

74LS Series Library Part 3 – 74LS(3).SLB				
74LS646	74LS652	74LS668	74LS669	74LS670
74LS673	74LS674	74LS682	74LS683	74LS684
74LS685	74LS686	74LS687	74LS688	74LS689
74LS716	74LS748	74LS795	74LS796	74LS797
74LS798	74LS848			

74S Series Library – 74S.SLB				
74S00	74S02	74S04	74S05	74S08
74S10	74S11	74S30	74S32	74S38
74S74	74S86	74S112	74S113	74S132
74S133	74S157	74S163	74S175	74S240
74S244	74S374	74S373		

Memory Devices Library – MEM.SLB				
2150	2210P	2212P	2716	27C16
2732	27C32	27C64	2764	27F64
2816	2817A	2864B	2864A	4164
SIL4256A	SIL4256C	4256	SIL4256D	SIL4256B
4416	48Z02	48T02	6116A	6164C
6168	6264A	6264	6267	7C128
9306	27C101	27C102	27128	27C128
27C256	27F256	27256	27C512C	27C512
27512	28C256	41264	43256C	50464
74S409	424256	511000		

Microprocessor Library Part 1 – Micro1.SLB				
5380	6502	6522	6800	68000
68008	6802	6809	6809E	6821
6850	8080A	8085	8086(MAX)	8086(MIN)
8087	8088(MAX)	8088(MIN)	8089	8237A
8251	8255	8259A	9516	9517A
9519A	9900	99105	9980A	M8719
S9901				

Microprocessor Library Part 2 – Micro2.SLB				
2650A	2651	2653	2655	2661
2672	2681	68010	68488	6852
6854	6875	6887	8031	8039
8155	8156	8212	8216	8224
8253	9902	S2350	S6551	S6810
S6840	Z80	Z80-CTC	Z80-DART	Z80-DMA
Z80-PIO	Z80-S10/0	Z80-SI0/2	Z8001	Z8002
Z8016	Z8030	Z8031	Z8036	Z8530
Z8531	Z8536	Z8671	Z8681	

PAL Library – PAL.SLB				
10H8	10L8	12H6	12L6	14H4
14L4	16H2	16L2	16C1	16L8
16P8	18P8	16R4	16R6	16R8
16RP4	16RP6	16RP8	16X4	16A4
16RA8	16V8	23S8	6L16	8L14
12L10	14L8	16L6	18L4	20L2
20C1	20L8	20L10	20S10	22P10
22XP10	20R4	20R6	20R8	20X4
20X8	20X10	20RS4	20RS8	20RS10
20RP4	20RP6	20RP8	20RP10	20XRP4
20XRP6	20XRP8	20XRP10	20RA10	22RX8
22V10	29M16	29MA16	32VX10	32R16

Surface mount components libraries

The Part I libraries follow the American National Standard ANSI/IPC-SM-782, and the libraries are identified by the use of a single letter M in the titles. The standard is by now out of date, and Part II libraries, using the double MM titles, are practical variants which allow more copper area, less critical placement and more tolerances all round. In addition, the MM libraries contain some components that are not contained in the original standard. Parts with the F suffix are flipped, with silk-screen on layer 9 for use on the bottom of the board.

PART I: ANSI / IPC-SM-782 LAND PATTERNS

Library File – SM.LLB				
RESISTORS				
0805	0305F	1206	1206F	1210
1210F				
CAPACITORS				
C1210	C1210F	C1812	C1812F	C1825
C1825F	C3216	C3216F	C3518	C3518F
C3527	C3527F	C3528	C3528F	C6032
C6032F	C7227	C7227F	C7243	C7243F
C7257	C7257F			
INDUCTORS				
L-A	L-AF	L-B	L-BF	
CYLINDRICAL PARTS				
MELF	MELFF	MLL34	MLL34F	MLL41
MLL41F				
SEMICONDUCTORS				
SOD80	SOD8OF	SOT23	SOT23F	SOT89
SOT89F	SOT143	SOT143F		

Library File – DSM.LLB

STANDARD PARTS (DUAL IN LINE)

8	8F	14	14F	16
16F	18	18F	20	20F
22	22F	24W	24WF	28
28F				

SMALL OUTLINE DEVICES

SO8	SO8F	SO14	SO14F	SO16
SO16F	SOL14	SOL14F	SOL16	SOL16F
SOL20	SOL20F	SOL24	SOL24F	SOL28
SOL28F				

SMALL OUTLINE 'J' LEAD DEVICES

SOJ14	SOJ14F	SOJ16	SOJ16F	SOJ18
SOJ18F	SOJ20	SOJ20F	SOJ22	SOJ22F
SOJ24	SOJ24F	SOJ26	SOJ26F	SOJ28
SOJ28F				

DIL FLATPACKS

MO003-10	MO003-10F	MO003-14	MO003-14F	MO004-10
MO004-10	MO004-10F	MO004-14	MO004-14F	MO004-16
MO004-16F	MO018-40	MO018-40F	MO019-24	MO019-24F
MO019-28	MO019-28F	MO020-36	MO020-36F	MO020-40
MO020-40F	MO021-16	MO021-16F	MO021-24	MO021-24F
MO021-36	MO021-36F	MO022-20	MO022-20F	MO022-42
MO022-42F	MO023-36	MO023-36F	MO023-50	MO023-50F

Library File – QSM(l).LLB

QUAD FLAT PACKS

QFP44	OFP44F	QFP48	QFP48F	QFP52
QFP52F	QFP54	QFP54F	QFP64	QFP64F
QFP70	QFP70F	QFP80	QFP80F	QFP100
QFP100F				

JEDEC TYPE 'A' DEVICES

JA28	JA28F	JA44	JA44F	JA52
JA52F	JA68	JA68F		

JEDEC 'J' LEAD PLASTIC DEVICES (PLCCs)

JJ20	JJ20F	JJ28	JJ28F	JJ44
JJ44F	JJ52	JJ52F	JJ68	JJ68F
JJ84	JJ84F	JJ100	JJ100F	JJ124
JJ124F				

JEDEC RECTANGULAR 'J' LEAD PLASTIC DEVICES

JJR18	JJR18F	JJR18L	JJR18LF	JJR22
JJR22F	JJR28	JJR28F	JJR32	JJR32F

JEDEC CERAMIC 'J' LEAD DEVICES

CJJ68	CJJ68F	CJJ84	CJJ84F

Library File – QSM(2).LLB				
JEDEC 'C' LEAD DEVICES				
JC28	JC28F	JC44	JC44F	JC52
JC52F	JC68	JC68F	JC84	JC84F
JC100	JC100F	JC124	JC124F	JC156
JC156F				
MINATURE PLCCs				
MPLCC84	MPLCC84F	MPLCC100	MPLCC100F	MPLCC132
MPLCC132F	MPLCC164	MPLCC164F	MPLCC196	MPLCC196F
MPLCC244	MPLCC244F			

Library File – QSM(3).LLB				
LEADLESS CHIP CARRIERS				
LCC16	LCC16F	LCC20	LCC20F	LCC24
LCC24F	LCC28	LCC28F	LCC44	LCC44F
LCC52	LCC52F	LCC68	LCC68F	LCC84
LCC84F	LCC100	LCC100F	LCC124	LCC124F
LCC156	LCC156F			

Library File – CSM(l).LLB

DIN 41612 STANDARD CONNECTORS

32AP	32APF	32AS	32ASF	64ABP
64ABPF	64ABS	64ABSF	64ACP	64ACPF
64ACS	64ACSF	96P	96PF	96S
96SF				

MOD 4 CONNECTORS

MOD4-6	MOD4-6F	MOD4-8	MOD4-8F	MOD4-10
MOD4-10F	MOD4-12	MOD4-12F	MOD4-14	MOD4-14F
MOD4-16	MOD4-16F	MOD4-20	MOD4-20F	MOD4-24
MOD4-24F	MOD4-26	MOD4-26F	MOD4-30	MOD4-30F
MOD4-32	MOD4-32F	MOD4-34	MOD4-34F	MOD4-40
MOD4-40F	MOD4-48	MOD4-48F	MOD4-50	MOD4-50F
MOD4-56	MOD4-56F	MOD4-60	MOD4-60F	MOD4-64
MOD4-64F				

IDC HEADERS

IDC10H	IDC10HF	IDC10V	IDC10VF	IDC14H
IDC14HF	IDC14V	IDC14VF	IDC16H	IDC16HF
IDC16V	IDC16VF	IDC20H	IDC20HF	IDC20V
IDC20VF	IDC26H	IDC26HF	IDC26V	IDC26VF
IDC30H	IDC30HF	IDC30V	IDC30VF	IDC34H
IDC34HF	IDC34V	IDC34VF	IDC40H	IDC40HF
IDC40V	IDC40VF	IDC50H	IDC50HF	IDC50V
IDC50VF	IDC60H	IDC60HF	IDC60V	IDC60VF

Library File – CSM(2).LLB				
SURFACE MOUNT DIL SOCKETS				
DIP6	DIP6F	DIP8	DIP8F	DIP14
DIP14F	DIP16	DIP16F	DIP18	DIP18F
DIP20	DIP20F	DIP22	DIP22F	DIP224
DIP224F	DIP24	DIP24F	DIP244	DIP244F
DIP246	DIP246F	DIP286	DIP286F	DIP32
DIP32F	DIP40	DIP40F	DIP48	DIP48F
DIP489	DIP489F	DIP64	DIP64F	
MISC. CONNECTORS				
SDL	SDLF	LOK	LOKF	
JUMPER / SHUNT ARRAYS				
SHUNT5	SHUNT5F	SHUNT6	SHUNT6F	SHUNT7
SHUNT7F	SHUNT8	SHUNT8F	SHUNT9	SHUNT9F
SHUNT10	SHUNT10F			
SWITCHES				
SW6	SW6F	SW8	SW8F	SW10
SW10F	SW12	SW12F	SW16	SW16F
SW18	SW18F	SW20	SW20F	SW24
SW24F				
CHIP CARRIER SOCKETS				
CC44	CC44F	CC52	CC52F	CC68
CC68F	CC84	CC84F	CC100	CC100F

PART II: COMMONLY USED PATTERNS MODIFIED FOR EASE OF MANUFACTURE

Library File – SMM.LLB				
RESISTORS				
0805	0805F	1206	1206F	1210
1210F				
CAPACITORS				
1812	1812F	C3216	C3216F	C3528
C3528F	C6032	C6032F	C7243	C7243F
SEMICONDUCTORS				
SOT23	SOT23F			
TRIMMERS				
POTA	POTAF	POTAH	POTAHF	POTB
POTBF				
BRIDGE RECTIFIER				
BR	BRF			

Library File – DSMM.LLB				
SMALL OUTLINE DEVICES				
SO8	SO8F	SO14	SO14F	SO16
SO16F	SO18	SO18F	SO20	SO20F
SO22	SO22F	SO24	SO24F	SOL20
SOL20F	SOL22	SOL22F	SOL24	SOL24F
SOL28	SOL28F	SOW24	SOW24F	SOW28
SOW28F				

Library File – QSMM(l).LLB

QUAD FLAT PACKS

QFP32	OFP32F	OFP44	QFP44F	QFP48
QFP48F	QFP52	QFP52F	QFP54	QFP54F
QFP56	QFP56F	QFP64	QFP64F	QFP64S
QFP64SF	QFP70	QFP70F	QFP80	QFP80F
QFP80S	QFP80SF	QFP100	QFP100F	

Library File – QSMM(2).LLB

JEDEC 'J' LEAD PLASTIC DEVICES (PLCCs)

JJ20	JJ20F	JJ28	JJ28F	JJ44
JJ44F	JJ52	JJ52F	JJ68	JJ68F
JJ84	JJ84F	JJ100	JJ100F	JJ124
JJl24F				

JEDEC RECTANGULAR 'J' LEAD PLASTIC DEVICES

JJR18	JJR18F	JJR18L	JJR18LF	JJR22
JJR22F	JJR28	JJR28F	JJR32	JJR32F

Appendix G: Using a plotter

At one time, the purchase of a plotter was something that could be contemplated only by a commercial or educational user of EASY-PC, but prices of plotters have dropped considerably in the last few years, encouraged by the excellent results that can now be obtained from low-cost ink-jet printers. One very considerable advantage of using a plotter is that plotting in several colours is easy, with no compromises in output quality as are experienced when some colour ink-jet machines are used. The other advantages include very high quality, with thinner lines than can reliably be produced by printers, and the ability to use a wide range of pen sizes.

Plotting from EASY-PC is done by way of the auxiliary program EASYPLOT which can be used from the DOS prompt or from within Windows. Your file (schematic or layout) must have been saved, and you will need to know the filename and directory.

When you start EASYPLOT by typing its name and pressing RETURN you will be asked for the name of the file that you want to plot. You must use the full filename with extension because it would be impossible otherwise to distinguish files that used the same main name for the SCH and PCB versions. After the file has been loaded in, EASYPLOT will report the size of the file and the number of tracks, pads, text items and symbols that are used. You will also get a message about whether or not the plot will fit the paper – this can be adjusted later. You can now alter the menu options using letter keys.

A: Alters output destination, using the parallel port LPT1, serial port COM1, or a file (you will be asked for a filename when you plot if you take this option). NOTE: To use LPT2 or COM2, select File then type the port name when you are prompted for a filename.

- Note that the HP 7575 plotter uses slightly different resolution in the two plotting directions – you should check the driver that is being used to find if this is taken into account. The Roland DXY plotter can use the HP7550A driver.

B: Specifies the fill type for solid lines and shapes. You can select Artwork, which uses a dense fill, Draft which is less dense (and therefore faster) or None (in which case some items are drawn in outline only.

C: Outlines. This also can use Artwork, Draft or None, and it determines how smoothly outlines, particularly sloping lines, are drawn.

D: Pen size, ranging from 0.1 mm to 0.9 mm in 0.05 mm steps. The size entered here must correspond to the pen that you are using, and this in turn must be small enough to deal with the smallest parts of the drawing.

E: Pen speed, ranging from 20 mm/second to 500 mm/second. This is set to the maximum speed that your pen can deal with, and you will need to consult the plotter manual to find what can be used. As a rough guide, use 500 for nylon tip, 100 for ceramic tip, 500 for roller-ball, 250 for OHP (spirit) pens, 250 for refillable wet-ink pens, 500 for disposable wet-ink pens and 500 for auto-feed pencils. These are maximum speeds and they may need to be reduced depending on the type of paper, the age of the pen, temperature, humidity and other factors.

F: Output scale, ranging from 0.24 to 4. This determines the scale of plot, and is more useful for schematics than for PCB work, unless the PCB has been produces to a scale other than 1:1. You may need to use this if the PCB or schematic cannot be plotted on the paper being used.

G: Print area. This is specified by entering the X and Y co-ordinates of the bottom left and the top right corners of the board or schematic that is to be plotted. You can also use this to specify a limited area to be plotted.

H: Maximum plot size. This is used when you are first using EASYPLOT, or when you are changing to a different plotter. You need to enter the co-ordinates of lower left and upper right corners of the *plotter area* in HPGL units. The dimensions can be obtained from the plotter manual. Some examples are provided in the EASY-PC manual. You cannot use EASYPLOT unless this option has been used. You can usually assume that 1" = 1016 HPGL units and 1 mm = 40 HPGL units.

I: Fill or Avoid pad holes. Use Avoid for single-sided boards unless a solder-resist plot is being made. Use Fill for all other boards.

J: Plot offset. Enter X and Y offset values to shift the plotted area. This can be used to show different layers offset.

K: Plot layer. You can enter a number 0 - 9 here for PCB layer. It is ignored when you are processing schematics.

L: Start plot.

M: Change name of input file – this is used following one plot to load in another file for the next plot.

N: Centre plot. This will centre the plotted schematic or layout on the paper.

O: Pen number (1 to 8). Used to select the pen in a multi-pen plotter.

P: Pads only. This is used in conjunction with option I. If you are not filling holes, answer Yes or No. If holes are being filled, you will be required to enter the extra width of the solder resist mask in thousandths of an inch. The possible numbers following a Yes answer for this case are (0), (4), (8), (12) or (17).

Q: Quit to DOS (or to Windows if EASYPLOT was started from Windows).

R: Restore defaults from disk. This restores the set of defaults that was last saved, using the file PLOT.DEF. If you want to use the defaults that were originally supplied (which you may have changed), copy this file from the original distribution disk.

S: Save defaults. Will save the changes that you have made in this menu to the file PLOT.DEF for use in future.

Figure G.1 shows suggested RS-232 pin connections for both 25-pin and 9-pin connectors for serial plotter cables. The normal RS-232 pin uses are tabled below.

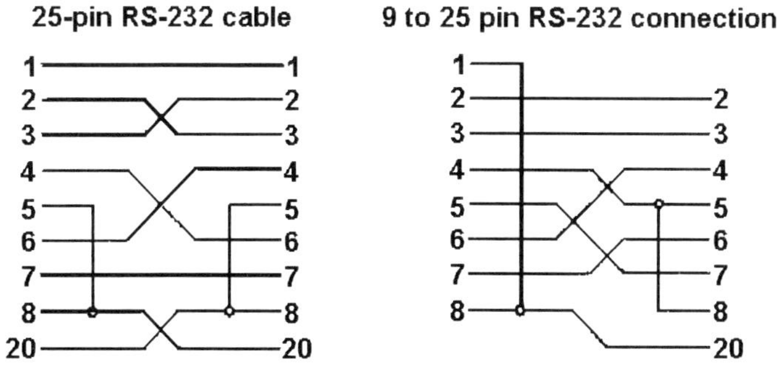

Figure G.1 The suggested RS-232 connections for driving a plotter. Your serial port must be set up for the correct baud rate, number of stop and start bits etc.

25-pin connector	
1	N.C. (not connected)
2	Transmit Data
3	Receive Data
4	Request To Send
5	Clear To Send
6	Data Set Ready
7	Ground
8	Carrier detect
9-19	N.C. (not connected)
20	Data Terminal Ready
22	Ring Indicator
23-25	N.C. (not connected)

9-pin connector	
1	DCD
2	RXD
3	TXD
4	Data Terminal Ready
5	Ground
6	Data Set Ready
7	Request To Send
8	Clear To Send
9	Ring Indicator

Index